The Dialectic
of Duration

Groundworks

Series Editors: Arne De Boever, California Institute of the Arts;
Bill Ross, Staffordshire University; Jon Roffe, University of New
South Wales; Ashley Woodward, University of Dundee

What are the hidden sources that determine the contemporary moment in continental thought? This series goes 'back to the source', publishing English translations of the hidden origins of our contemporary thought in order to better understand not only that thought, but also the world it seeks to understand. The series includes important French, German and Italian texts that form the lesser-known background to prominent work in contemporary continental philosophy. With an eye on the contemporary moment—on both world-historical events and critical trends—Groundworks seeks to recover foundational but forgotten texts and to produce a more profound engagement not only with the contemporary but also with the sources that have shaped it.

The Dialectic of Duration

Gaston Bachelard

Translated and annotated by
Mary McAllester Jones

Introduction by
Cristina Chimisso

ROWMAN & LITTLEFIELD
INTERNATIONAL

London • New York

Published by Rowman & Littlefield International, Ltd.
Unit A, Whitacre Mews, 26-34 Stannary Street, London SE11 4AB
www.rowmaninternational.com

Rowman & Littlefield International, Ltd. is an affiliate of Rowman & Littlefield
4501 Forbes Boulevard, Suite 200, Lanham, Maryland 20706, USA
With additional offices in Boulder, New York, Toronto (Canada),
and Plymouth (UK)
www.rowman.com

This edition first published by Rowman & Littlefield International, 2016
Translation © Mary McAllester Jones 2000
Introduction © Cristina Chimisso 2000

Published in French by Presses Universitaires de France as *La Dialectique
de la durée*
© Presses Universitaires de France, 1950
Bibliothèque de philosophie contemporaine
108 boulevard Saint-Germain, 75006, Paris

First English translation published by Clinamen Press, 2000

A slightly adapted section of this work previously published in
Jones, Mary McAllester, Gaston *Bachelard, Subversive Humanist* © 1991.
Reprinted by permission of the University of Wisconsin Press.

British Library Cataloguing in Publication Data
A catalogue record for this book is available from the British Library

ISBN: HB 978-1-78660-058-5
     PB 978-1-78660-059-2

Library of Congress Cataloging-in-Publication Data

Library of Congress Control Number: 2016947516

∞™ The paper used in this publication meets the minimum requirements of
American National Standard for Information Sciences—Permanence of Paper for
Printed Library Materials, ANSI/NISO Z39.48-1992.

Printed in the United States of America

Contents

*Page references for the 2nd French edition (France: PUF, 1993) are
given in the margin*

Groundworks Series Preface

The purpose of the new Groundworks series is to present English translations of significant texts of European philosophy, with an emphasis on those works which enjoy an influence on the shape and evolution of continental thought which has been felt only at one remove in the Anglophone tradition. The present volume, Gaston Bachelard's *The Dialectic of Duration*, has remained little known in the Anglophone world despite its importance; it represents Bachelard's direct counterpoint to the work of his great contemporary, Henri Bergson. Both are pre-eminent names in the history of the French academy, whose work each in its way is testament to the suppleness and adaptability of a broadly phenomenological tradition in the hands of genuinely creative thinkers. Despite the fact that Bergson's reputation waned dramatically in the latter half of his career, this has been recuperated posthumously with equal force, not least due to the influence of his most famous reader, Gilles Deleuze. Bachelard's reputation suffered no such reversal; his presence has been an enduring feature of the landscape of continental thought, while his prodigious output has steadily worked its way into translation. It would be fair to say, though, that for readers coming to *The Dialectic of Duration* first in this translation, and perhaps more familiar with the themes of continuity, intuition and the virtual, as they have settled from Bergsonism into the Anglophone tradition, this of all Bachelard's work retains the greatest power to unsettle and to reframe the strands of thought which are in danger today of becoming too readily emblematic of continental thought as such. Yet for all that, there is a great deal to this book which reaches beyond the disputed territory between the two philosophers; it was in this work that Bachelard first marshalled all the components of his visionary philosophy of science, with its steady insistence on the human context and subtle encompassing of the irrational within the rational. As such it represents a privileged point of entry to the work of one of the great modern philosophers of science.

Translator's Note

In his discussion of time in *The Dialectic of Duration*, Gaston Bachelard
tackles a wide range of topics and in doing so, faces the translator with
a number of challenges. In translating this book my aim has been—as
always—fidelity both to Bachelard's French and to the English language; conflicts of loyalty do of course arise and in such instances the
translator's first duty must, in my view, be to Bachelard. His ideas are
stimulating, thought-provoking and sometimes difficult: smoothing out
these difficulties would be a disservice to Bachelard. I have therefore
sought to retain the sense one has while reading the French text that he
is grappling with new ideas, working his way towards understanding
them through language and style that in places reflect their complexity.
Bachelard's delight in language is also very evident in this book: his
choice of unexpected, striking words is often illuminating, helping our
understanding, while his fondness for word-play and neologism forces
us to attend and to think. The translation of these neologisms offers a
particular challenge: I have added notes where I think these may baffle
English readers, while seeking to retain in the translation the slight
shock of Bachelard's neologisms to the French ear.

The range of material Bachelard covers in *The Dialectic of Duration*
has stimulated much discussion with colleagues and friends, and I wish
to thank them warmly for the time they have given me and for their
interest. As someone who has specialised in Bachelard for many years,
often in face of the puzzlement of those more caught up with philosophical and literary fashion, I have been very pleased—surprised even—by

the interest in Bachelard so readily shown by those with whom I have discussed him in the course of this translation. Discussions of Bachelard with my students over the years have shown how thought-provoking he remains, with his capacity for opening to us entirely new perspectives; I would like to thank them—and in particular those members of my class of '99 so smitten with Bachelard—for their enthusiasm.

I am indebted to the University of Wisconsin Press for kind permission to use here translated extracts from *The Dialectic of Duration* that first appeared in my book *Gaston Bachelard, Subversive Humanist*, published by them in 1991. Some small amendments have been made here to those original translations.

This translation owes much to the support of my husband Robert McAllester Jones: for his understanding, his patient proofreading and his discussions, I am deeply grateful.

November 1999
Mary McAllester Jones

a life of seclusion, or with strengthening moral solitude. Let everyone take, each in their own way, their first steps along the road leading to the pool of Siloam[1] and the very sources of the person! Let everyone free themselves, as they themselves choose, from the contingent stimuli that draw them out of themselves! It is in the impersonal part of the person that the philosopher must discover areas of repose and reasons for repose, with which he will make a philosophical system of repose. Through philosophical thought, being will free itself from a life force that carries it far from individual goals and expends itself in imitated actions. When the intelligence has been restored to its speculative function, it will be revealed to us as a function that creates and strengthens leisure. Pure consciousness will be revealed as the capacity for waiting and for watchfulness, as the freedom and the will to do nothing.

We were thus led quite naturally to examine the negating powers of the mind. We first examined this negation at its root, recognising that the mind could conflict with life, oppose ingrained habits, and in a way make time flow back on itself so as to bring about renewals of being, returns to original conditions. Why did we not regard the negative and positive actions of time as equally important? Our intention was to go straight to the metaphysical heart of the problem and so what had to be founded was a dialectic of being in duration. Now, as soon as we had acquired through meditation some skill in emptying lived time of all its excess and ordering the different levels of temporal phenomena into a series, we realised that these phenomena did not all have the same *duration:* the conception of one single time carrying our soul away for ever and ever along with things can only correspond to an overview which offers an inadequate summary of the temporal diversity of phenomena. Botanists who limited their science to saying that all flowers fade would just be doing the same thing as some philosophers who underpin their theories by repeating that all things pass away and that time flies. We very soon saw that between this passing of things and the abstract passing of time there is no synchronism, and that temporal phenomena must each be studied according to its appropriate rhythm and from a particular point of view. When we examined this phenomenology in its contexture, taking it at any one of its levels but with the proviso that our study be restricted to this same level, we saw that it always comprises a duality of events and intervals. In short, when we looked at it in the detail of its flow, we always saw a precise, concrete duration that teemed with lacunae.

Our first task had to be to establish metaphysically—against the Bergsonian thesis of duration—the existence of these lacunae in duration. We thus had to begin by discussing Bergson's famous discourse on the idea of nothingness and set about restoring the balance between the passage of being to nothingness and of nothingness to being. This was an indispensable basis for founding the alternative of repose and action.

This is not in our view a pointless debate since by drawing on a dialectical conception of duration we can, as we have set out to show in this book, help solve the problems posed by psychological causality or, more accurately, by psychological causalities. As we examine layer by layer the different sequential levels of the psyche, we see the discontinuities of psychic production. If there is continuity, it is never at the level specifically under examination. For example, 'continuity' in the effectiveness of intellectual motives does not lie at the intellectual level; it is *postulated* at the levels of emotions, of instincts and of interests. Psychic concatenations are therefore often hypotheses. In short, in our opinion psychic continuity poses a problem and it seems to us impossible not to recognise the need to base complex life on a plurality of durations that have neither the same rhythm nor the same solidity in their sequence, nor the same power of continuity.

Were we able to convey to the reader our firm belief that *psychic continuity is not given but made,* we would of course still have to show how duration is constructed and how the various permanencies of being are grounded in its diverse attributes.

We have been encouraged in this difficult task by a number of ideas, first and foremost by those of Gaston Roupnel, whose lessons drawn from life were dispensed along the paths of Burgundy, amidst the vineyards.[2] There, in this landscape shaped by human beings, Roupnel showed us that different things and different times slowly adjust to each other, that space acts on time and time reacts upon space. Ploughed fields depict figures of duration every bit as clearly as figures of space; they show us the rhythm of human toil. A furrow is the temporal axis of work and evening repose is the field's boundary-mark. How poorly are such temporal moulds as these expressed by a continuously, regularly flowing duration! How much more real, as the basis of temporal effectiveness, must seem the concept of rhythm!

As Roupnel also asks, what remains of the historical past, what lasts from it? Only that which has reasons for beginning again. Thus, alongside duration through things, there is duration through reason. The same is always true: all true duration is essentially polymorphous; the real action of time requires the richness of coincidence and the syntony of rhythmic efforts. We are beings with a strong constitution, living in sure and certain repose, only if we know how to live according to our own rhythm by rediscovering each in our own way, in moments of weariness and despair, the momentum of our origins. This is what the lovely story of Siloam illustrates, teaching us how to restore our former soul courageously, deliberately and rationally. We have already devoted a book[3] to this myth and so shall not return to it. It has, however, influenced our thinking so profoundly that it had to be mentioned here, at the beginning of this new book.

Since what has most duration is what is best at starting itself up all over again, we came to understand that the concept of rhythm is the fundamental concept of time. We were thus led to put forward an argument that may seem very paradoxical but that we shall attempt to justify. It is this: the phenomena of duration are constructed by rhythms, rhythms that are by no means necessarily grounded on an entirely uniform and regular time. Ideas found in the books of Maurice Emmanuel, Lionel Landry and Pius Servien has helped us formulate this argument briefly and succinctly here. It was precisely because they had no metaphysical aims that these books were chosen to support a metaphysical argument. They seemed to us better able to help us show the essentially metaphorical character of the continuity of temporal phenomena. To have duration, we must entrust ourselves to rhythms—that is to say, to systems of instants. Exceptional events must find resonance in us if they are to mark us deeply. In the end then, we would venture to turn the platitude 'life is harmony' into a truth. Without harmony, without a well-ordered dialectic, without rhythm, life and thought cannot be stable and secure: repose is a happy vibration.[4] Lastly, we were a few years ago given the opportunity to read an important work which, as far as we know, has never been published commercially and which has the splendidly luminous and thought-provoking title *La Rythmanalysis*.[5] Reading this convinced us that in exactly the same way that we refer to psychoanalysis, so there is a place for rhythmanalysis in psychology. A sick soul—especially one that suffers the pain of time and of despair—has to be cured by living and thinking rhythmically, by rhythmic attentiveness and rhythmic repose. It must first be cleared of all false permanence and ill-made durations, and disorganised *temporally*. In the days of people like Novalis, Jean-Paul Richter and Lavater, it was the fashion to disorganise psyches that were stuck in contingent forms of sentimentality and therefore unable to lead aesthetic and moral lives.[6] In our view though, this kind of disorganisation at the emotional level is still too crude. Here again, we have tried to take our philosophy of negativity a stop further and bring our desire to dissociate to bear on the very fabric of time itself. Ill-made rhythms have been tossed into delirium, forced ones given peace and listless ones aroused; syntheses of being have been sought in the syntony of becoming, and finally to all life that has the quiet, ordered movement of a wave new vigour has been imparted by the soft tones of intellectual liberty. Sometimes, in moments of happiness that are all too rarely found, we rediscovered more natural rhythms, both simpler and more peaceful. We would emerge from these rhythmanalysis sessions feeling more serene. Our repose would be more cheerful, more spiritual and poetic, as we lived this well-ordered temporal diversity. Ill-prepared as we might be for such emotions by our all too abstract culture, we felt that these rhythmanalytical meditations brought us a kind of philosophical echo of the joys of poetry. Between pure thought and pure poetry we would suddenly find links, agreements, even Baudelairian correspondences. We moved not just from one meaning to another but from meaning to the soul. Could it then be that poetry is not an accident, a detail, a mere diversion of being? Could it be the very principle of creative evolution? Do human beings have a poetic destiny? Could it be that they are on this earth in order to sing the dialectic of joy and pain? These are questions of a kind we are not qualified to go into. We have therefore reduced our task to the minimum and summarised Pinheiro dos Santos's most essential ideas in a short concluding chapter here, turning them slightly in the direction of an idealist philosophy in which the rhythm of ideas and songs would gradually command the rhythm of things.

NOTES

1. Bachelard refers obliquely here to Gaston Roupnel's *Siloë* (Paris: Delamain et Boutelleau, 1927). Bachelard's first book on time was a response

Introduction

'Of Bergsonism we accept everything but continuity' (p. 20). With this remarkable sentence, Bachelard concludes his somewhat ironic presentation of Bergson's philosophy in chapter one of *The Dialectic of Duration*. Bachelard explicitly presents his book as a critique of one of the most popular philosophies of the first half of the Twentieth Century: that of Henri Bergson. The above claim leaves no doubt that this is an anti-Bergsonian book, for to deny continuity is to reject Bergsonism. In fact the reader does not need to reach chapter one to realise the polemical character of Bachelard's book; the cover would suffice, for the title openly reveals the author's intentions. Dialectic cannot but destroy Bergson's *duration*, which is continuous and devoid of contradictions.

Bachelard was implacable: indeed, four years earlier, in 1932, he had already published a book aimed at a radical criticism of Bergson's philosophy: *L'intuition de l'instant*.[1] This title was fashioned to have the same rhetorical effect as the title *The Dialectic of Duration* was to have; just as the latter subverts one of the fundamental principles of Bergson's philosophy, duration, the former contradicts the other, intuition. For Bergson, intuition enables us to grasp 'real time', or duration. The time perceived by intuition, being continuous, is qualitatively different from the time of physics, which is made up of instants. Physics cannot go beyond its discontinuous conception of time precisely because it is not intuitive knowledge. An intuition of the instant, as Bachelard's title suggests, would create a bridge between those elements that Bergson intends to separate: intuition and instant. Intuition is for Bergson the organ of philosophical knowledge, while the instant is a scientific concept which exists precisely because intuition is not available to science. Science relies on intellect; Bergson concedes that the intellect serves to grasp one level of reality, namely the realm of inert matter. He claims that:

> The intellect [is] a special function of the mind, essentially turned toward inert matter.[2]

Bergson does not regard scientific knowledge as relative, despite its being intellectual knowledge; rather, he believes that it can reach an absolute status provided it applies to its proper object. However, when it departs from inert matter, it becomes less and less reliable. Biology for him is already a borderline case, for its object is constituted by living beings.[3] He thinks that in order to grasp life, intellect is not adequate, and that therefore science will always fail in this endeavour.

> The intellect is characterised by a natural inability to comprehend life.[4]

For Bergson, in order to comprehend life, intuition is needed, and philosophy is the discipline that rests on it. He vindicates a role for philosophy which is distinct from that of the 'positive sciences', in its method and its object; its method being intuition and its object life. Despite his concessions to science, philosophy appears to be not only different from science but also superior to it, for it succeeds where science fails:

> Chemistry and physics will [n]ever give us the key to life.[5]

In science objects are conceived as static, because intellectual knowledge is only capable of this much: '*of immobility alone does the intellect form a clear idea*'.[6] Becoming is therefore for science only a succession of immobile pictures. Bergson calls scientific and ordinary knowledge 'cinematographic',[7] for, as in moving pictures, the movement is an illusion created by a succession of static frames. In Bergson's eyes, philosophers have not managed to go beyond this fragmentation: the ancient philosophers denied the importance of time, and sought truth outside it. Modern philosophers have considered time, but they have shared the world-view of the sciences, and have reduced it to a succession of snapshots. Indeed, time has been modelled on space and equated to it, most notably by Kant. For Bergson 'real time', being continuous, escapes the intellect, for '*of the discontinuous alone does the intellect form a clear idea*'.[8]

The provocation of Bachelard's first anti-Bergsonian book, *L'intuition de l'instant*, is apparent: for Bergson intuition is the faculty which allows us to go beyond the fragmented time—instants—in order to access the 'real' continuous time, duration. *The Dialectic of Duration* expresses the same criticism from the opposite end: to dialecticise duration amounts to reducing it to the discontinuous time of physics. Dialectic of duration is not a small correction of Bergsonism, but rather its negation. And yet, Bachelard claims that his aim is to correct Bergsonism:

> We wish therefore to develop a discontinuous Bergsonism, showing the need to arithmetise Bergsonian duration so as to give it more fluidity, more numbers, and also more accuracy in the correspondence the phenomena of thought exhibit between themselves and the quantum characteristics of reality (p. 20).

If read as a mere discussion of Bergson's philosophy, *The Dialectic of Duration* is rather paradoxical, as shown by Bachelard's aim at 'arithmetising' it. However, this book is much more than a comment on Bergsonism; it is the exposition of Bachelard's own conception of time, history and knowledge. Why do it by opposing Bergson?

One reason is inherent in Bachelard's own view that advancement of knowledge proceeds dialectically. A new doctrine, or a new scientific theory, cannot be formulated without a polemical discussion with an existent, and competing, doctrine. Indeed, Bachelard argues that it is time to substitute Kant's 'architectonic reason' with 'polemical reason'.[9] In *Le nouvel esprit scientifique* (1934),[10] he had already set out his epistemology of oppositions: he had interpreted Einstein's physics as an opposition to classical mechanics, and had argued that modern epistemology opposed Cartesianism. In *La philosophie du non*, published in 1940, he was going to provide an articulated theoretical defence of polemical reason. In this book, he argued that in science:

> Above all we must recognise the fact that new experience says *no* to old experience, otherwise we are quite evidently not up against a new experience at all.[11]

Philosophy, in order to proceed, must say *no* to old philosophy, and to competing philosophies. In *The Dialectic of Duration*, philosophy is made to develop by saying *no* to Bergsonism.

The other reason for Bachelard to direct his attack on Bergson is more circumstantial, bound up with the competing currents of philosophical thought and practice in the French universities at the time. In the first half of the Twentieth Century, Bergsonism had a tremendous impact on Parisian intellectual life. Bergson's lectures at the Collège de France were crowded beyond belief; in illustrations of the time, we see people climbing windows to get a glimpse of the celebrated philosopher. This philosopher declared the sciences limited in their object and in their scope. He proposed a new role for philosophy, freed from subordination to the sciences. At the time of *The Dialectic of Duration*, Bachelard was Professor of Philosophy at the University of Dijon, but four years later returned to the Sorbonne, where he had obtained his PhD, as Professor of History and Philosophy of the Sciences, and as Director of the Institute of History of the Sciences and Techniques. This Institute had been created in 1932 by Abel Rey, Bachelard's supervisor on one of his two doctoral dissertations. Bachelard's philosophy, like those of his mentors at the Sorbonne, represented a defence of scientific rationalism, to which Bergson was a formidable threat. In 1936, Bergson was seventy-eight, and had long left his chair at the Collège de France. However, his philosophy was still very influential; his last book, *La pensée et le mouvant*, had been published in 1934, and he had received the Nobel prize for literature in 1927. What made his philosophy more menacing for the Sorbonne philosophers of science was that it was kept alive precisely in the domain of philosophy of science. Bergson's chair had gone to his follower Edouard Le Roy. Le Roy, a mathematician by training and a militant Catholic, regarded himself as a disciple of Bergson and of Poincaré, and intended to combine their philosophies. His examination of the possibilities and limits of science brought him to conclude that scientific knowledge is an essentially conventional, if useful, practice. Above science, he posed intuitive knowledge, which grasps reality directly and truly, as Bergson had taught. Bachelard's other mentor at the Sorbonne, the historian of philosophy Léon Brunschvicg, had done battle with Le Roy's 'new philosophy' in the pages of the *Revue de Metaphysique et Morale* already at the turn of the century. Brunschvicg had defended 'intellectualism' against Le Roy's attacks. Bachelard went much further in this defence, and his two anti-Bergsonian books constitute the outline of his version of rationalism.

Rather than undermining Bergsonism from within, Bachelard challenges it from a different perspective; that of rationalism. Bachelard does not try, as, for instance, Bertrand Russell did, to show that Bergson misunderstood the concept of number as mathematics assumes it, or to demonstrate that Bergson's solution of Zeno's paradox of Achilles and the tortoise is fallacious.[12] The project of *The Dialectic of Duration* is much more ample than finding inconsistencies or logical problems with Bergson's argument. It is part of the building of a new philosophy, which comprises not only an epistemology but also a historiography and an ethics.

By opposing Bergson's project of divorcing philosophy from science, Bachelard defends the view that 'science in fact creates philosophy' as he declares in *Le nouvel esprit scientifique* [*The New Scientific Spirit*].[13] However, by this Bachelard does not mean that the role of philosophy is simply to accept the results of the sciences and to adopt their language. It suffices to read a few pages of *The Dialectic of Duration* to be assured that we are rather far from any project of scientistic or positivistic philosophy. For Bachelard what science has to teach philosophy is of a general order; first of all, its method, but not the specific methods of the specific sciences, which for him are diverse and open to change. Rather, Bachelard thinks that philosophy should adopt the dialectical method that characterizes scientific inquiry. Science, he argues throughout his extensive output, is always the result of a fertile clash of new experiences and projects against old ones; it is always the result of interaction and dialogue between individuals; finally, it is a positive knowledge which comes out as negation of other possibilities and other solutions.

Not surprisingly, Bachelard attacks Bergson's conception of intuition as the way of attaining true knowledge. For Bachelard, no knowledge is possible without mediations and rectifications. The criticism of immediate knowledge is a constant theme in Bachelard's work, starting with one of his two doctoral dissertations, *Essai sur la connaissance approchée*. In the chapter 'Intuition et réalisme en mathématiques', he judges mathematical intuition as very often unable to lead 'to a rigorous knowledge'.[14] For him, intuition 'hinders ... the freedom of the spirit'.[15] One could say that Bachelard holds an 'anti-Rousseauian' philosophy: for him, human nature is not naturally good, and certainly the human mind is not: it is not able to know if untrained. In order to attain scientific knowledge, the mind has to fight against 'easy intuitions'. Bachelard returns many times to the necessity of revising 'first knowledge' or 'first intuitions', for they are generally wrong.[16] The 'first experience', the experience which has not undergone a critical examination, is the first of the obstacles to objective knowledge that he analyses in *La formation de l'esprit scientifique*.[17] For Bachelard, science is a model of knowledge precisely because it advances by revising and correcting previous achievements. In his vision of history of science, a new way of thinking, which he calls 'the new scientific mind', emerged with non-Euclidean geometry as the overcoming of the immediate, intuitive and naive approach of Euclidean geometry. What is intuitive for Bachelard is the product of imagination and desire and this hinders the course of science. Moreover, for Bachelard scientific activity represents a model also because it requires interactions between individuals; modern physics, he argues, cannot be carried out by isolated scientists. Indeed, he claims that 'all rationalism is interrationalism'.[18] By contrast, intuitive knowledge, which Bergson defends, is private.

Bachelard's conception of knowledge as discursive, indeed as polemical, requires negations. A positive item of knowledge emerges against something that is *not* it. In the first chapter of *The Dialectic of Duration*, Bachelard considers Bergson's criticism of negation and of the idea of nothing. Bergson regards the idea of nothing as a pseudo-problem, despite its persistence in the history of philosophy. Nothing cannot be imagined or perceived; it is only a name, even though it plays a social and pedagogical role. Consequently, for Bergson negation is in fact an affirmation:

> Thus, whenever I add a 'not' to an affirmation, whenever I deny, I perform two acts: (1) I interest myself in what one of my fellow- men affirms, or in what he was going to say, or in what might have been said by another *Me*, whom I anticipate; (2) I announce that some other affirmation, whose content I do not specify, will have to be substituted for the one I find before me. Now, in neither of these two acts is there anything but affirmation.[19]

As a result, for Bergson there is no disorder: any negation of order is simply the affirmation of another order. There are different *kinds* of order, rather than disorder. Bergson sees qualitative differences where Bachelard sees oppositions. For Bergson, philosophy, through its organ, intuition, is able to grasp qualitative differences whereas science reduces everything to quantities and formal models. However, he argues that philosophers so far have not pursued the possibilities of their discipline. In particular they have failed to consider the qualitative distinctions of the various processes which go under the name of 'becoming'. In Bergson's words:

> That which goes from yellow to green is not like that which goes from green to blue: they are different *qualitative* movements. That which goes from flower to fruit is not like that which goes from larva to nymph and from nymph to perfect insect: they are different *evolutionary* movements. The action of eating or drinking is not like the action of fighting: they are different *extensive* movements. And these three kinds of movements themselves—qualitative, evolutionary, extensive—differ profoundly.[20]

For Bergson the perception of a discontinuity in a process is due only to the limitation of our comprehension. He wants to show that we cannot stop perceiving; that suppressing something only means replacing it with something else; and that negating something means affirming something else. This is why Bachelard calls Bergson's philosophy 'a philosophy of fullness' and his psychology 'a psychology of plenitude'. He polemically remarks:

> This psychology is so rich, so multifarious and mobile[21] that it cannot be contradicted. ... In these conditions, life cannot go in fear of some total failure. If intellect dims then instinct awakens. Indeed human beings—who in dedicating themselves to intellect have put so much at risk—have at least held on to sufficient instincts to keep themselves in ignorance and error. They will take decisions, clearly and rationally, but in between these they move with all the confidence of a sleepwalker (p. 15).

When Bachelard wrote *The Dialectic of Duration*, Bergson had already attacked the dominant school of the experimental psychology of his time, in *Matter and Memory* first and then in several talks, notably in 'Le cerveau et la pensée: une illusion philosophique' and in 'L'âme et le corps'.[22] In particular, he had criticised the doctrine of parallelism, according to which to any psychic state there corresponds a neurological state. The adherents to parallelism denied any distinction between brain and mind. Bergson proposed his own psychology which restated a dualism between body and soul, matter and consciousness. Matter—our body—acts, whereas the mind remembers. In its autonomous experience, the mind does not experience those temporal discontinuities and voids which characterise the perception of material things. As a result the mind is capable of preserving the past in its entirety, so that it always remains part of the present.

Pierre Janet, Professor of Experimental Psychology at the Collège de France, criticised Bergson's conception of time and memory in his course of lectures delivered in 1927–1928 and immediately published.[23] Bachelard relied on Janet's criticism, for it was aimed to show the psychological implausibility of duration as formulated by Bergson. In chapter 2 of *The Dialectic of Duration*, Bachelard extensively quotes from Janet's lectures and pays particular attention to his 'psychology of commencement'. The importance Janet attaches to the beginning and end of actions and events lends support to Bachelard's theory of time as discontinuous. In his own words:

> we need the concept of the instantaneous in order to understand the psychology of beginning (p. 48).

Moreover, Janet's insistence on the human and dialectical character of memory was particularly useful to Bachelard's goal. In the lecture titled 'Le récit', Janet characterises the act of memory as social and as being aimed at 'triumphing over an absence'. Indeed human beings use memory to 'fight against absence'. Bachelard endorsed Janet's conception of memory as something to be constructed. Bachelard wanted to show that our personal history is neither memory of a continuity nor contains the entirety of our past, but rather selected memories. For Bachelard, our personal history is the account of our actions. Its continuity is not a reality apprehended by an immediate intuition, but it is rather a construction realised through the reasons we attach to our distinct actions (p. 42). For him, the way in which we know our past is not different from the way we know material objects, for both require reflection and judgement. Consequently, Bachelard rejects Bergson's attempt to make intellectual knowledge a particular type of knowledge the object of which cannot be life:

> In vain do we try to differentiate between understanding a process and living it (p. 42).

For Bachelard the knowledge of our history is equally mediated. He cites Pierre Janet expressing a concept which was going to be a key to his own philosophy: to know is always to teach. Knowledge is always a dialectical exchange and a discussion, even in the case of solitary knowledge, in which the discussion becomes internal to the individual. Knowledge of something is realised when it becomes transmittable. This requires an order. We evaluate past events, select and organise them according to a plan:

It is impossible to know time without judging it (p. 43).

In *The Dialectic of Duration*, Bachelard presents his concept of 'rational memory' or 'rationalised memory', which he fully develops in his later books (p. 76).[24] Rational memory reconstructs the past according to present conceptions and finalities. This is for him the way the sciences use their past, and he calls this process 'recurrent history'. The recurrent history of a science is constituted by a selective choice of theories interpreted in the light of present theories. Bachelard claims that science is able to incorporate previous theories, despite the fact that it advances by contradicting them, precisely because it reconstructs its history in this way. In other words, only a reinterpretation of the past can create the continuity of history of science, which otherwise would exhibit ruptures and contradictions. Memory as a coherent narrative is for Bachelard the result of a construction rather than a given, and so is any 'psychic continuity' (p. xiii).

A fundamental point of Bachelard's criticism is that Bergson's philosophy prevents human beings from judging, choosing and acting according to a plan. Bergson's creative evolution proceeds thanks to an internal, 'natural' motor: the vital impulse. By contrast, for Bachelard, change (rather than 'evolution') is brought about by work, commitment and critical assessment of past ideas and practices. That *The Dialectic of Duration* was intended to provide not only a metaphysical justification of discontinuity but also an ethical reason for it is clearly stated.

The concluding chapter of the book, 'Rhythmanalysis',[25] is an articulated proposal of how we should give a 'rhythm' to our life. This rhythm is analogous to the rhythm of intellectual knowledge, that is to say, to its dialectic process. Just as knowledge proceeds by alternating negation and creation, our life should be an alternation of the

effort of intellectual knowledge and the rest from its demands. Indeed
Bachelard presents this book as an 'introduction to the teaching of
a philosophy of repose' (p. xi). This repose, during which imagination can express itself, should also receive a 'rhythm': continuity for
Bachelard is just not creative. He argues that poetic rhythm has been
achieved by Surrealism, an avant-garde movement that he greatly
admired. In chapter 7 of *The Dialectic of Duration*, he explains how
surrealist poetry provides a good example of temporal dialectic. If
it is not understood by psychologists, logicians and literary critics,
Bachelard goes on, it is because they want to judge it by superimposing on it the schemes of continuity, without allowing for the dialectical freedom on which it is constructed. In 'Surrationalism', an article
published the same year as *The Dialectic of Duration*, Bachelard
compares the rhythm of surrealist poetry with the rhythm he proposes
for rational knowledge. He suggests that the dialectic between rational
plot and dream devised by Tristan Tzara for poetry could represent a
model for rational knowledge, so that 'rationalism' would turn into
'surrationalism'.

For Bachelard the alternation of these two dialectical activities—
intellectual and imaginative—should form the rhythm of our own life.
Indeed he proposes the dialectical play of presence and absence of
intellectual work as a philosophy of life. The dialectic of activity and
rest put forward in *The Dialectic of Duration* is reflected in the subjects
of Bachelard's books. Some of them are about science while others are
about the works of imagination. In all of them, he supports an alternation
of intellectual and imaginative activities which would form a 'complete
anthropology'. His conception of the interplay between intellect and
imagination is indeed 'polemical'. Intellectual knowledge must purify
itself from imagination, and it advances by overcoming the obstacles
that imagination poses by diverting the attention from its proper object
and spoiling the logic of intellectual knowledge. Conversely, imagination must purify itself from the constraints of intellect in order freely to
express itself. In his view, neither faculty can be eliminated, because the
atrophy of one would stop the activity of the other, and because human
beings need to exercise both.

Far from being a merely polemical book, *The Dialectic of Duration*
contains all the major themes of Bachelard's philosophy of science
as well as of his philosophy of imagination. He continued his ambi-
tious project of providing a basis for a 'complete anthropology' in his

subsequent books, which are often more specific and focused than *The Dialectic of Duration* but develop the ideas exposed in this short but still all-encompassing work.²⁶

Cristina Chimisso

NOTES

1. *L'intuition de l'instant* (Paris: Stock, 1992 [1931]). Translated as *Intuition of the Instant* (US: Northwestern University Press, 2013).
2. Henri Bergson, *Creative Evolution* (Mineola, NY: Dover, 1998 [1907]), 206.
3. Henri Bergson, *The Creative Mind* (New York: Hubner, 1946 [1903–1923]), 43–50.
4. Henri Bergson, *Creative Evolution* (Mineola, NY: Dover, 1998 [1907]), 165.
5. *Ibid.*, 31.
6. *Ibid.*, 155.
7. *Ibid.*, 306.
8. *Ibid.*, 154.
9. Gaston Bachelard, *L'engagement rationaliste* (Paris: Presses Universitaires de France, 1972), 'La psychologie de la raison': 27–34.
10. Paris, Presses Universitaires de France, 1991 (Engl. trans.: *The New Scientific Spirit* [Boston: Beacon Press, 1984]).
11. Gaston Bachelard, *La philosophie du non. Essai d'une philosophie du nouvel esprit scientifique* (Paris: Presses Universitaires de France, 1988 [1940]), 9 (Engl. trans.: *The Philosophy of No. A Philosophy of the New Scientific Mind* [New York: The Orion Press, 1968], 9).
12. Bertrand Russell, *The Philosophy of Bergson* (Cambridge: Bowes and Bowes, 1914).
13. Gaston Bachelard, *Le nouvel esprit scientifique* (Paris: Presses Universitaires de France, 1984 [1934]), 7 (Engl. trans.: *The New Scientific Spirit* [Boston: Beacon Press, 1984], 3).
14. Gaston Bachelard, *Essai sur la connaissance approché* (Paris: Vrin, 1987 [1927]), 170.
15. *Ibid.*, 169.
16. Gaston Bachelard, *Le nouvel esprit scientifique* (Paris: Presses Universitaires de France, 1984 [1934]), 40–41 (Engl. trans., 38–39).
17. Gaston Bachelard, *La formation de l'esprit scientifique: contribution à une psychanalyse de la connaissance objective* (Paris: Vrin, 1993 [1938]), ch. 1.

18. Gaston Bachelard, *Le rationalisme appliqué* (Paris: Presses Universitaires de France, 1986 [1949]), 21.

19. Henri Bergson, *Creative Evolution* (Mineola, NY: Dover, 1998 [1907]), 289.

20. *Ibid.*, 304.

21. On the following page, Bachelard says, 'Bergsonism that has been accused of a predilection for mobility'. Bachelard here refers to Julien Benda, who had attacked Bergson in his *Le Bergsonisme ou une philosophie de la mobilité* (Paris: Mercure de France, 3rd ed., 1912). Ironically, he was going to extend his attack to other philosophers, including Bachelard, as their work become available. In *De quelques constantes de l'esprit humain. Critique du mobilisme contemporain (Bergson, Brunschvicg, Boutroux, Le Roy, Bachelard, Rougier)* (Paris: Gallimard, 1950), he published new and old essays against the above philosophers and others such as Abel Rey and Benedetto Croce, because in his eyes they conceived knowledge as something never fixed, but rather always in evolution.

22. *Matière et memoire,* first published in 1896, was constantly reprinted. The two talks, delivered in 1904 and in 1912, respectively, were published together with other talks in Bergson: *L'énergie spirituelle. Essais et conférences*, Paris 1919.

23. Pierre Janet, *L'évolution de la mémoire et de la notion du temps. Compte-rendu intégral des conférences d'après les notes sténographiques* (Paris: Chahine, 1928).

24. Bachelard develops the concept of rational memory in particular in *Le rationalisme appliqué* (Paris: Presses Universitaires de France, 1986 [1949]), ch.1; and in *L'activité rationaliste de la physique contemporaine* (Paris: Presses Universitaires de France, 1951), ch.1. In *Le rationalisme appliqué*, Bachelard counterposes 'rational memory' to 'empirical memory' and argues that the former, as 'memory of reason [and] of coordinated ideas', follows laws which are different from those of empirical memory (p. 2).

25. This is based on the 'rhythmanalysis' proposed by a certain Lucio Pinheiro dos Santos.

26. For a bibliography of Bachelard, see: H. Choe, *Gaston Bachelard. Epistemologie. Bibliographie* (Frankfurt am Main: Lang, 1994). For works on Bachelard in English, see: Cristina Chimisso, *Gaston Bachelard: Critic of Science and the Imagination* (UK: Routledge, 2001); Mary Tiles, *Bachelard: Science and Objectivity* (UK: Cambridge University Press, 1984); Mary McAllester Jones, *Gaston Bachelard, Subversive Humanist: Texts and Readings* (US: University of Wisconsin Press, 1991): Joanne H. Stroud, *Gaston Bachelard: An Elemental Reverie of the World's Stuff* (US: Dallas Institute Publications, 2015); Zbigniew Kotowicz, *Gaston Bachelard: A Philosophy of the Surreal* (UK: Edinburgh University Press, 2016). For Bergson and Bergsonism:

P. A. Y. Gunter, *Henri Bergson: a bibliography,* Rev. 2nd ed. (Bowling Green, Ohio: Bowling Green State University, 1986); Gilles Deleuze, *Bergsonism* (New York: Zone Books, 1988); R. C. Grogin, *The Bergsonian Controversy in France, 1900–1914* (US: University of Calgary Press, 1988); Leonard Lawlor, *The Challenge of Bergsonism* (US: Continuum, 2003).

Chapter 1

Relaxation and Nothingness[1]

'Who will tell me how, all through existence, my whole person
has been preserved? What was it that carried me, inert, full of
life and spirit, from one end of nothingness to the other?'

Paul Valéry, *A.B.C.*

I

Bergson's philosophy is a philosophy of fullness and his psychology   1
is a psychology of plenitude. This psychology is so rich, so multifari-
ous and mobile that it cannot be contradicted; it makes repose active
and functions permanent; it can always draw on so many things that
the psychological scene will never be empty and success also will be
ensured. In these conditions, life cannot go in fear of some total failure.
If intellect dims then instinct awakens. Indeed human beings—who in
dedicating themselves to intellect have put so much at risk—have at
least held on to sufficient instincts to keep themselves in ignorance and
error. They will take decisions, clearly and rationally, but in between
these they move with all the confidence of a sleepwalker. They move
even faster when they do not know where they are going, when they
entrust themselves to the life force that carries humankind along, when
they turn away from personal solitude. Thus, our life is so full that it is
active when we do nothing. In a way, there is always something behind

15

us, Life behind our life, the life force beneath our impulses. Our whole
past is also waiting there behind our present, and it is because the self   2
is old and deep, rich and full that it can truly act. Its originality stems
from its origin, not a matter of some chance discovery but of memory.
We are bound to ourselves, our present action neither disconnected nor
gratuitous but always of necessity expressing our self, just as a quality
expresses a substance. In this respect, Bergsonism has the facility of
all substantialism and the ease and charm of all doctrines of interiority.

While Bergson would doubtless refrain from inscribing the past in
matter, he does in fact inscribe the present in the past. Thus, the soul
is seen to be a thing behind the flux of its phenomena; it is not really
contemporary with its own fluidity. And the Bergsonism that has been
accused of a predilection for mobility has not however set itself within
duration's very fluidity. It has maintained the solidarity of past and
future and also the viscosity of duration, with the result that the past
remains the substance of the present or to put in another way, that the
present instant is never anything other than the phenomenon of the
past. And thus it is that in Bergsonian psychology, duration that is
full and deep, continuous and rich serves as the spirit's substance. In
no circumstances can the soul separate itself from time; like all that
are happy in this world, it is always possessed by what it possesses.
Ceasing to flow would mean ceasing to subsist; were we to leave this
world's ways we would leave life. When we come to a standstill we
die. Thus, while we think we have broken with the idea of the soul as
substance, we are in fact shaping our innermost being from the stuff
of an indestructible duration. Panpsychism is now nothing other than
a panchronism, and the continuity of thinking substance nothing other
than that of temporal substance. Time is alive and life is temporal.
Before Bergson, the equation of being and becoming had never been
so well established.

However, as we shall see at greater length in due course, for Bergsonism the creative value of becoming is limited by the very fact of   3
fundamental continuity. Time has to be left to take its course if it is to
do its work. In particular, the present can do nothing. Since the present carries out the past just as a pupil carries out a task the teacher
has set, the present can create nothing. It cannot add being to being.
Here again, Bergsonism has developed by following the intuition of
fullness. For this school of thinkers, the dialectic always goes directly
from being to being without nothingness being brought in between

them. Jankélévitch has in fact suggested that the famous disquisition on nothingness be regarded as the basis of Bergson's philosophy. We know that for Bergson the idea of nothingness is in the end richer than that of being, simply because this idea would only be brought in and only become clear if an additional function of annihilation were added to the different functions by which being is established and described. In Bergson's view therefore, the idea of nothingness is functionally richer than the idea of being. Thus, with regard to the knowledge we have of them, no substance could have any void and no melody be broken by an absolute silence. The substance we know must always express itself. In a way, all the possibilities of human thought and action inevitably become attributes of the substance under consideration if we bear in mind an ingenious theory of negative attribution. May we in fact subsequently come to deny a quality we once attributed to substance? If so, we are expressing an error we have made rather than a deficiency of substance. Thus conceived as a sum of possibilities, substance is inexhaustible. The possible as a possible can never fall because it remains possible; in the same way, the probable that has been clearly seen to be such will always keep its precise value. The possible and the probable therefore have perfect continuity and it is for this reason that they are very precisely the spiritual attributes of substance as it presents itself to our analysis in the problem of knowledge.   4
The significance of Bergson's subtle critique can only be understood if we carefully take up position on idealist ground where knowledge of being is concerned, and avoid descending too quickly to the ontological domain. We shall then see the full importance of problematic judgement. From this point of view, the possible is a memory and a hope. It is what we once knew and hope to find again. It is thus well suited to filling in at least the discontinuities in knowledge of being, though maybe not the gaps in being. In this way is the unending dialogue of mind and things prepared, and the continuous fabric woven that lets us feel substance within us, at the level of our innermost intuition, despite the contradictions of external experience. When I do not recognise reality it is because I am absorbed by memories that reality itself has imprinted in me, because I have returned to myself. For Bergson, there is no wavering, no interplay, and no interruption in the alternative we have between knowledge of our innermost self and of the external world. I act or I think; I am a thing or a philosopher. And through this very contradiction, I am continuous.

The same comments would be made, Bergson argues, about the psychology of a decrease in psychological intensity as about the psychology of annihilation, since in his view the impression that an intensity decreases while still remaining comparable to itself is as artificial and misleading as the idea one might have of absolute nothingness. For Bergson, when there is a decrease there is a change in nature. Spiritual substance is thus enfolded in endless, prodigiously diverse attributes, and all the degrees of attribution have an equal force of attribution. The subtleties of psychological analysis have a charm that can immediately be ranked among the soul's riches. Bergson ascribes the emotion generated by his subtle psychological analysis to the fundamental value of our feelings. For him, nuance is a colour. The impression is given then 5 that the Bergsonian soul cannot stop feeling and thinking, that feelings and ideas are endlessly renewed on its surface, glinting in the flow of duration like a river's water in sunshine.

The impression of plenitude given by Bergsonian psychology can indeed be increased by the perfectly complementary character of some opposition. Not only is the absence of a form automatically the presence of a different form, but the lack of a function is sure to activate a function that is the very opposite of what was done initially and failed. It would seem that without this immediate rectification of one function by another, being is no longer of use to itself. An essential failure would shatter being, breaking up the becoming that is totally inseparable from being. Failure must thus remain partial, superficial, and rectifiable. It must not prevent the continuous and deep success of being. This in fact metaphysical success is so certain that failure in one respect is more than compensated by success in another. In the general theory of the life force, there is an entire argument for ontological compensations that justifies, for the individual and above all for the species, the most unfortunate initiatives. There is nothing more Bergsonian than this idea of the plurality of different means to reach the same goal, a plurality that gives all curiosity, all testing and seeking out, an undisputed positive value. Life is never absolutely and unconditionally at risk. Indeed Bergson, with his subtle analysis of the risk which gave rise to intellect, always maintained that this risk was active under the pressure of circumstance, in the struggle for life, as it kept hold of the past as something stable and sure, desiring repose, security and peace, in the secret ambition of being to acquire more duration. He always maintained that behind intellect, 6 instinct stood guard. Should instinct fail, torpor would be there to keep watch, so to speak, as a positive function of the psyche that can oblige being to wait without destroying it. Coming back to the life force and its bold innovations, Bergson has no doubt shown very clearly that the greatest success is to be found where risk is the highest, but once again risk, in his view, has an aim and a function, in other words it has a history, a development, a logic, a myriad empirical and rational guarantees that found the continuity of the most adventurous of lives.

All these arguments do not however deal with the metaphysical essence of risk and Bergson has written nothing on and for risk, nothing on absolute and total risk, risk that has no aim and no reason; he has not written about the strange emotional game that leads us to destroy our security, our happiness and our love, nor about the sense of exaltation that draws us to danger, newness, death, and nothingness. As a result of this, the philosophy of the life force has been unable to give full meaning to what we shall call the purely ontological success of being, that is to say to the constant re-creation of being by itself, in the spiritual act of consciousness in its purely gratuitous form, as resistance to the call of suicide and triumph over the charms of nothingness. Bergsonism positioned itself systematically in relation to the evolution of species; the individual's free act, whose meaning and place it had however better shown than any other school of philosophy, was somehow eliminated in the whole evolution of the species. Lastly, in Bergsonism the free act seems to lack that purely intellectual causality that binds but does not constrain: it remains an accident. The theory of creative evolution, based on purely biological evolution, on evolution that is long, obscure and tenacious, thus discounted anything corresponding to the will to destroy or struggle just for the sake of struggle. It first and foremost attributed to being a continuum of growth, to the species continual life through seeds, and to living destiny a force that of necessity would never end, since any interruption breaks up a force even more surely than it does a thing. Bergson's thought is therefore always and every- 7 where guided by the same fundamental idea: being, movement, space, and duration cannot receive lacunae; they cannot be denied by nothingness, by repose, by points, by instants; or at least such negations are condemned to remain indirect and verbal, superficial and ephemeral.

To sum up, whether it be in our intuition of duration, our conceptions of being, or even in the service of our functions, according to Bergsonism we are given over to an *immediate*, *deep* continuity which can only superficially be broken by what lies outside, by appearance, and by language that claims to describe it. Discontinuities, fragmentation, and negation only appear to be devices which help exposition; psychologically, they are in the expression of thought and not in any way within the psyche. Bergson did not attempt to activate dialectic at the level of existence nor even at that of intuitive, deep knowledge; he considered that dialectic did not go beyond the dialogue of the soul with reality, and that the experience that goes from things to the self was an interplay of images which kept a fundamental homogeneity.

This then, in our view, is how the metaphysical connection between non-being and being can briefly be characterised in Bergsonism. We shall now go on to criticise this school of thought on one specific point. Since any critique is clarified by its end-point, let us say straightaway that of Bergsonism we accept everything but continuity. Indeed to be even more precise, let us say that from our point of view also continuity—or continuities—can be presented as characteristics of the psyche, characteristics that cannot however be regarded as complete, solid, or constant. They have to be constructed. They have to be maintained. Consequently, we do not in the end see the continuity of duration as an immediate datum but as a problem. We wish therefore to develop a discontinuous Bergsonism, showing the need to arithmetise Bergsonian duration so as to give it more fluidity, more numbers, and also more accuracy in the correspondence the phenomena of thought exhibit between themselves and the quantum characteristics of reality.

8

II

There is no doubt that our critique must first be brought to bear on the order of discourse and with reference in fact to Bergsonian proof. After this, we can proceed to positive psychological investigations; we shall then enquire whether Bergsonism has given their rightful place to psychological negativism, coercion, and inhibition. Once we have made this detailed study of the psychology of annihilation, we shall attempt to establish that annihilation implies nothingness as its limit, in exactly the same way that qualification implies substance as its support. From the functional standpoint that we shall adopt, it will be seen that there is nothing more normal or more necessary than going to the limit and establishing the relaxation of function, the repose of function, the non-functioning of function, since function must obviously often stop functioning. It is at this point that we shall see the interest of taking the principle of negation back to its source in temporal reality itself. We shall see that there is a fundamental heterogeneity at the very heart of lived, active, creative duration, and that in order to know or use time well, we must activate the rhythm of creation and destruction, of work and repose. Only idleness is homogeneous; we retain only what we reconquer; we maintain only what we resume. Moreover, simply from   9 the methodological point of view, it will always be in our interests to establish a connection between the dialectic of different entities and the fundamental dialectic of being and non-being. We shall therefore bring philosophy back to this dialectic of being and nothingness, convinced moreover that it was not by some accident of history that the first Greek philosophers were led to this problem. Pure thought must begin by refusing life. The first clear thought is the thought of nothingness.

Where discourse is concerned, Bergson's argument in *L'Evolution créatrice* comes down to saying that there are no truly negative actions and that, as a result, negative words only have meaning in terms of the positive words they negate, all action and all experience being unfailingly first expressed positively. Now it seems to us that the priority thus given to the *positive misrepresents* the perfect correlation between words when they are, as they have to be, translated into the language of action. A concept is formed by experience and analysed by actions. It is for this reason that, for example, the word *empty*, taking its meaning from the verb *to empty*, can be said to correspond to a positive action. A well-trained intuition would thus conclude that emptiness is simply the disappearance, whether imagined or real, of a particular substance without it ever being possible to speak of a direct intuition of emptiness. All absence would thus be consciousness of departure. Such, in the end, is Bergson's argument. Now, while it is very true that you can only empty what you first found full, it is just as accurate to say that you can only fill what you first found empty. If the study of fullness is to be clear and rich, it must always be a more or less detailed account of filling something. In short, there seems to us to be a perfect correlation between emptiness and fullness. One is not clear without the other and in particular, one idea is not clarified without the other. If the   10 intuition of emptiness is refused us, then we have the right to refuse that of fullness.

Bergson has recently expressed an objection to the facile clarity of intellectual methods but has not convinced us.[2] We see the relations between intuition and intellect as more complex than a simple opposition. We see them as constantly co-operating when they come into play. There are intuitions at the root of our concepts: these intuitions are unclear: they are wrongly thought to be natural and rich. There are intuitions too in the way we put concepts together; these essentially secondary intuitions are clearer: they are wrongly thought to be artificial and poor. Let us look briefly at the psychology of a scientific mind tormented by the idea of vacuum, of void and emptiness. Such scientists have read the long history of these ideas; they can use the difficult techniques of creating a vacuum, always concerned about the possibility of a micro-leak; they doubtless know how fallacious this idea is since suddenly, just when they thought they could define the emptiness of matter, they have seen it inhabited by radiation. They are therefore better prepared than anyone else to understand a theory according to which emptiness from one point of view is automatically fullness from another. But they are not satisfied with the idea of something happening automatically. They sense a new problem: they try or will try to create a vacuum from two combined points of view; they will attempt to eliminate both matter and radiation. Their concept of emptiness consequently becomes richer, more diverse, and because of this, more clear. Indeed no scientist will claim a priori clarity for their experimental ideas. They are as prudent as intuitionist philosophers, whose patience they share. We can moreover bring these two groups together in equal esteem: as Bergson has rightly said, a philosophical intuition requires long contemplation. Such difficult contemplation as this, which has to be learned and which no doubt could be taught, is not far from being a discursive method of intuition. This is what gives us grounds for attaching, as of prime importance, the psychology of the clarification of ideas to the logical definition of these ideas. A balance can therefore be established between the inverse conceptualisation of emptiness and fullness and we can bring into balance, not though as starting points but as summarising factors, the contrary concepts of fullness and emptiness.

There is naturally the same detailed, discursive correlation set up between being and nothingness when we really try to live the dialectical oscillation of making and annihilating. If we say we are doing so on the basis of a logical dialectic, an immediate dialectic, with being and nothingness regarded straightaway as ready made *things*, we shall be

11

open to Bergson's criticism. There is indeed a shocking lack of balance between these two ideas taken as substitutes for two realities! Is it not glaringly obvious that nothingness cannot be a *thing*? That repose cannot be a kind of movement? Is it not equally obvious that being is an asset we have realised, the most solid, stable thing there is?

We shall not however let ourselves get involved in an a priori choice and shall keep bringing our opponents back to the need for being to be established, by them as well, by stages and discursively. By what right could being be established in one go, beyond and above experience? We call for complete ontological proof, for the discursive proof of being, for detailed ontological experience. We want to touch both the wounds and the hand. The miracle of being is as extraordinary as that of resurrection. We shall be no more content with a sign in order to believe in reality than are our opponents with a failure in order to believe in the ruin of being. Indeed, this ontologising requirement will be made the heart of our polemic. We believe moreover that we are thus placing the problem where it ought to be posed: is not knowledge essentially a polemic?

12

III

When Bergson compares the two judgements 'this table is white', 'this table is not white', he underlines on the one hand the immediate and determinate character of the first and on the other, the indirect, indeterminate character of the second. He thus presents the second judgement as marked by a verbal polemic and condemned to remaining powerless in face of the decisive first intuition. In our view, however, all the values of verification must be transmuted, the power of conviction being given above all to negative judgements. In other words, all effective judgements—that is to say all judgements that engage the consciousness—are for us negative judgements; they are the decisive arguments in a fierce polemic. It is not in fact a matter of *repeating* that the table is white but of *discovering* or *letting others discover* that the table is white. You cannot really hope to carry out a productive psychological investigation if you take an example in which the impression studied does not provoke discussion. Let us not therefore take our examples in those limp affirmations born of habit and associated with

lazy recollections. Let us try to grasp the mind in its essential act of
judgement.

Will you then take a judgement of discovery? Have you discovered
a blue dahlia? Do you therefore declare that this flower is despite this a
dahlia? This is an admission that previously you imagined that for this
flower such colouring was impossible. Your judgement of discovery, of
surprise and exclamation, is thus no more immediate and direct than   13
any negative judgement. It was preceded by the inverse judgement, by
the flimsy, irrational inverse belief that there is no such thing as a blue
dahlia...

Will you now take an affirmative judgement that expresses for you
something that you have long known? It is very certain that this judgement is only a psychological act if it is peremptory; it must not be mur-
mured half-heartedly or caught up in the prattle of reminiscence. Do not
forget that we are dealing with *proofs* of being, or better with proofs of
the real connection of being to itself; it is Being, objective being as well
as subjective being, it is your being and your whole reason that you are
engaging in this discussion. For discussion there is since you are affirming forcefully; and since you are expending nervous energy, a little of
your living soul and duration, it means that something or someone is
obstructing you: you are contradicted and so you affirm.

But is it perhaps in solitude that you pursue your thoughts, your affirmations seeming to you full and tranquil, strong and fundamental? This
is because you score a cheap victory over the possible adversary that
you always do in fact imagine in order to personify the initial negation.
Back in his prison after having renounced his 'errors', Galileo murmurs 'yet it does move'. He murmurs this in suffering, in the rancour
of defeat, in a stifled polemic. But all his thought is a reaction against
previous official denials.

Let us also go into the heart of a stubborn child; make this child
be quiet; make him, make her, recant their desire and this desire will
return, reinforced by resistance, fuelled by negation, in a soft, tenacious
affirmative judgement. Always and everywhere, we only affirm psychologically what has been denied or what we conceive to be deniable.
Negation is the nebula out of which real positive judgement is formed.

There would perhaps be a method for legitimising the primacy of
positive judgement but it would not be very Bergsonian at all, since
it would be based on a kind of logical necessity: knowledge could   14
be said to have to begin by affirmations and express, in affirmative

forms, ingenuous first impressions. Indeed, this argument amounts to a departure from concrete psychology, from psychology that has proofs. Scientific psychology can in fact no more invoke a first impression than astronomy can rely on the Book of Genesis. We neither think with our first impressions, love with our original sensibility, nor want with our initial and substantive will. There is the same distance between our childhood and our present self as between dream and action. After all, the wonder of first thought is perhaps grounded on preceding doubt that is all the more methodical for being more natural. Truth suddenly appears against a background of errors, and singularity against one of monotony; temptation too is seen against a background of indifference, and affirmation against one of negations. When affirmation comes to have a psychological sense, it is because it is reacting against negations or previous ignorance. Its energy is a function of the number and importance of the negations it challenges.

To sum up, affirmation is by no means synonymous with positive knowledge. It by no means has the privilege of fullness and certainty. It is a mistake to see it as immediate and primary. We cannot agree with Bergson when he wishes to unbalance the dialectic of positive and negative judgements by in a way filling thought with affirmative values that are themselves full and complete. We would prefer to break the balance in the opposite way, struck as we are by the negating value of all knowledge that is truly actual. In effect, psychological life must be grasped in its acts, in its flow, and not in its source which is always hypothetical and slight. All knowledge taken at the moment of its constitution is polemical knowledge; it must first destroy in order to make room for its constructions. Destruction is often total and construction never completed. The only clear positivity of knowledge there is can be found in consciousness of necessary rectifications, in the joy of imposing an idea. And without going as far as the polemical principle of knowledge, the whole psychology of insinuation, persuasion, and polite discussion could show us the same waves of dialectical thought, though slower and more gentle now. Here again, we must patiently sketch in a blurred background for positive, clear thought. Schopenhauer remarked ingeniously here that 'in order to get other people to accept our contradiction of their ideas, there is nothing more appropriate than the phrase "I used to think like that, but etc."'.[3] The better to contradict, we pretend to accept; we 'pick up' what they have said in order to prevent an unpleasant confrontation. Here we have an example of *continuity*

15

behaviour which does indeed underline the real discontinuity. Moreover in a feigned affirmative judgement, do we not see psychological negativism's greatest success? Were we to give it full affirmative value we would be fooled, imitating the learned ignorance of mathematics teachers as they act out for a moment their faith in the truly extraordinary hypotheses that will lead them to an absurd conclusion.

There is lastly one other, rather paradoxical way of contradicting Bergson's thesis, which is to generalise it. In effect, his proposal that a destructive thought be brought in to account for the very particular idea of *nothingness* seems to us to be what happens with every concept. The psychological significance of a specific concept cannot be better determined than by describing the process of conceptualisation in the course of which it was formed. Now this conceptualisation is the history of our refusals rather than of our assents. A clear concept must bear the trace of all that we have refused to incorporate into it. Generally speaking at the beginning of a conceptualisation, the vague and variable hues of a phenomenon must be erased so that its constant features can be drawn. All precise knowledge leads to the annihilation of appearances, to the   16 establishment of a hierarchy among phenomena, and the attribution to them in some way of coefficients of reality or if it is preferred, of coefficients of unreality.

Reality is thus analysed by means of negations. Thinking involves disregarding certain experiences and willingly casting them into the shadow of nothingness. If it is objected that the positive experiences we have erased still subsist, we shall reply that they subsist without playing a part in our present knowledge. We shall therefore now look at the problem from a functional point of view. We shall see that it is from a simply functional, no longer ontological point of view, that classification into affirmative and negative judgements has real psychological value.

IV

Certainly, concepts only have meaning when they are incorporated into judgements. This is a theory that has been developed at length by modem psychology, and we only need to draw its metaphysical conclusions. As Jean Wahl has put it, succinctly and subtly, 'as the mind moves towards more precision, it turns facts into factors'.[4] It is useless to attempt, by virtue of some kind of logical hierarchy of concepts, to place simple concepts endowed with intrinsic clarity in an immobile empyrean, with at their apex and in pride of place the concept of Being. The demand for precision is not satisfied by immediate clarity. Concepts multiply and diversify as they are applied, as they become factors of thought. Precise Being itself has to give us many proofs; we only accept it after a process of qualification that is diverse and mobile, tested and rectified. Thus, what *is* must psychologically *become*. We cannot think Being without linking to it a gnoseological becoming. If it is taken in its maximum synthesis, thought being must be an element of becoming. We shall try to show this functional element at the centre of action, at the centre of the verb.

17

Since our thought expresses virtual as well as real actions, its culmination is the very moment of decision. There is in particular no synchronism between thinking about acting and the real development of action. Contracting an action to fit the limits of the decisive instant thus constitutes both the unity and the absoluteness of that action. The movement will be completed in some way or other, determined as it is by subordinate, unsupervised mechanisms; for temporal behaviour, the essential thing is to begin the movement or better, to permit it to begin. All action is ours by virtue of such permission. Now this permission, which reflects action and is wholly conceived as the realisation of a possibility, develops in an atmosphere that is lighter than real action. Realisation is less opaque than reality. There is then, above lived time, thought time. This thought time is more aerial and free, more easily disrupted too and resumed. It is in this mathematised time that the inventions of being lie. It is in this time that a fact becomes a factor. We do not describe this time correctly by calling it abstract since it is in this time that thought acts and prepares the concrete manifestations of Being.

Permission to act can however be more easily centred than action itself. We therefore propose firstly to centre on the verb, the relations expressed by a judgement, rather than looking for their roots in either predicate or subject. We are, in our view, being faithful to Bergson's teaching here.[5] We shall next propose, at the centre of the verb, to reduce all action to its decisive, unitary aspect, which we can indeed take to be instantaneous if we compare it with real development, slow and multiple as this is. Here, we are breaking up Bergsonian continuity, preferring a hierarchy of instants. Far from language having its roots in the spatial aspect of things, its true mental function lies for us

18

in the temporal, ordered aspect of our actions. It is the expression of our preferences. We shall then go on to emphasise the ordering power of the life of the mind, following what Paul Valéry says in *Monsieur Teste* and stressing 'duration's delicate art, time, its distribution and its regulation—its expenditure on things it has well chosen, in order to nurture them particularly'. We shall thus see that the cohesion of our duration is made from the coherence of our choices and of the system coordinating our preferences. Yet this entire development will have no meaning unless we can at this stage isolate the very essence of the idea of *permission to act*. This permission is attached to the verb through the dialectic of *yes* and *no*. It appears to be added on and secondary to all theories of interiority which claim they can immediately grasp thought that is necessarily synchronous with life, rooted in life and proceeding at the same pace as life. This will not be the case with regard to theories that put forward the idea of thought freed from life, suspended above life and also able to suspend life. We shall then understand that all judgements are judged, and that it is this judging that prepares and gauges true psychological and biological causality. Exceptional decisions direct the thinking being's evolution. As regards judgement, affirmative or negative character is a functional and indeed essential addition. Thus the most peremptory, the most certain and most constant judgement is a conquest of fear, doubt, and error. As Von Hartmann well understood 'even the wish to remain in the present state assumes that this state can end and the fear that this possibility will be realised: we see a double negation here. Without the idea of ending, the wish to continue would not be possible'.[6] So thought proceeds: a *no* counters a *yes* and in particular a *yes* counters a *no*. An object's very unity is the result of our overall view of it, its diversity resulting from our refusal or our scattered concentration. Unity can never be given to an object unless it is grasped in the unity of an action, and our knowledge of an object can never be diversified unless the actions in which it is involved are multiplied, these actions being understood as separate. The schema of the temporal analysis of a complex action is necessarily a discontinuity.

There is in fact no other way of analysing an action than by beginning it all over again. And we have to begin it again by 'disassembling' it, that is to say by enumerating and ordering the decisions which constitute it. Moreover, it would be fanciful to attribute an essential role to the duration of an action that assembles. It would be futile to *lengthen verbs*

19

to enable them to be better understood, since this lengthening would in no way concern the verb's essential role. Saying that an action *has duration means* that we are still refusing to describe its detail. Were we to complete an analysis of an action which has duration, we would see that this analysis is expressed in separate phrases centred on instants of very precise singularity. From this point of view then, actions that assemble cannot be contiguous, and still less continuous. And what fragments thought is not the handling of solids in space but the dispersion of decisions in time. Once an action has been willed, once it is a conscious action and one that draws on reserves of psychic energy, it cannot flow continuously. It is preceded by hesitation, it is expected, deferred, provoked, and these subtle distinctions prove that it is isolated and that it appears in a dialectical wave-motion. Subsequently, when actions have to be bound together, we shall see the superiority of the mind to life in this respect; we shall see that, to preserve itself, life has no choice other than to fend off anything that would unbind it. The *wisdom of the function* will then be recognised. If we look to the agreement of successive functions for what binds life, rather than to a purely energetic propulsion, we shall soon accept the *reality of the order* of decisive instants. We shall be obliged to say that order is not in duration, that instead duration is indeed the affirmation of a useful, psychologically effective order. We can no doubt accept, as Bergson does, that in space disorder is but an order we had not foreseen, and that the dialectic of order and disorder has no spatial basis. But a temporal upheaval shatters life and thought, in both their detail and their principle. We die of an absurdity. Here, disorder is indeed a fact; it is a factor of nothingness. In order to think, to feel, to live, we must bring order into our actions by holding instants together through the reliability of rhythm, and by uniting reasons for coming to a vital conviction. But this is a point that we shall be looking at in detail later. For now, all we wish to do is to prepare our opposition to Bergson's argument that language is rooted in solids and that intellect learns its lessons from metric geometry. We shall then go on to try to bring out the realising value of order when order is taken to be a prime factor. It is therefore in *wise action* that we shall seek the principles of continuity.

20

V

Actions are not always positive and indeed from the standpoint of psychological action, where psychological functions are concerned, we can see there is a dialectic that again transposes the dialectic of being and nothingness.

Before we examine this functional dialectic, it is again necessary to show that for Bergson the constant action of functions corresponds to the fullness of being.

Indeed, from the psychological point of view, one is struck as one reads Bergson's work by the paucity of references which could provide the elements of an analysis of coercion and inhibition. The will is always positive and the will to live is, as in Schopenhauer, completely permanent. It is truly a force. Being wishes to create movement. It does not wish to create repose. 21

No doubt there are halts and there are failures; for Bergson though, the cause of failure is always external. It is matter that opposes life, colliding with life's momentum which it slows or deflects in a downward curve. If ever life could develop in some rarefied place, sustaining itself on essences that are the very staff of life, then it would rise in a single movement to its culmination. Life is thus broken or divided up when there is an obstacle to it. Life is a conflict in which you always have to be wily and always have to diverge along indirect paths. Such is the ancient image that came into being with *homo faber*, overwhelmed as he was by his tasks.

Yet if we take matter with the many obstacles it constantly puts in our path, matter around which we pick our way, that we assimilate and reject in our philosophical attempts to understand the world, then the question arises as to whether in Bergsonism matter really has enough characteristics to meet the often contradictory diversity of its functions. It seems not. On the contrary, one has the impression that for Bergson matter is identical to the failures it brings about. It is the substance of our disillusion, of our setbacks and mistakes. It is encountered after failure, never before. It substantialises repose after fatigue, and never repose that is finely constructed on a real point of equilibrium.

Why then not look at failure in itself, when reasons for acting are contradicted and a function which ought to act is non-functional? This would have given us an example of fundamental disorder, of a disorder of time and also of the mind.

Moreover, we have only to go into the psychology of hesitation to reveal the tissue of *yes's* and *no's*. Life opposes life, the body devours itself, and the soul gnaws itself away. It is not matter that is the obstacle. Things are but opportunities for us to be tempted; temptation is in us, it is both a moral and a rational contradiction. Fear too is in us, obviously present before there is danger. How would we understand danger without it? And the most insidious of worries is indeed the product of a lack of worry. As Schopenhauer said, it worries me when nothing worries me. We have only to dematerialise affectivity just a little to see it ripple in waves. 22

And if we dematerialise the problem of adaptation we shall come to the same conclusions. Indeed, if we take adaptation from the standpoint of the human psyche, in our attempts to become rational and knowledgeable beings, we see that it is distinct from the accidents of life. It is instead the fruit of curiosity, of meticulous care in bringing the harmony of being to completeness and in creating diversity in being. Yet for this very reason, curiosity is straightaway bordered by loss of interest: being wishes to change. The being that has succeeded has no taste for maintaining itself in that success. Curiosity grows dull, hopping hither and thither. And then the joy of discovery is countered by a kind of need to destroy, in a kind of inverse curiosity. We have only to point out this negating aspect of mental life for many biological and psychological characteristics to be better understood. We are aware of very many dark points that mark out all that would die in us, the shadow of Death as it were permeating Life. We understand why it should be that psychoanalysis has recently given great importance to the death instinct, to necrophilia, and to the need to lose that gives a new and very dialectical meaning to the need to gamble.

However, were all these psychological observations to appear secondary and ineffectual, and were it not seen that what plays over the surface of being reverberates down to its very principle, then we still have an argument in reserve that we consider decisive. Indeed, where physiology itself is concerned, the need for the non-functioning of function is so obvious, so natural, that it does not even occur to us to point it out. From the standpoint of energy, all functions are limited by thresholds of action. In vain are functions thought to be damped down, sleeping, or latent. Just slowing down is already a sign of discontinuity! If we take as our starting point the function in its complex action, we should in fact see that in slowing down it completely abandons certain of its characteristics. Indeed, this slowing is a descent right down a very 23

real staircase which is marked by many thresholds of differentiation.
On the lowest of its steps the most sharply distinct of all dialectics is
clearly in play, the law of *all or nothing* whose importance Rivers has
demonstrated in some detail in his book on the unconscious.[7]

VI

These brief notes will, we consider, be sufficient to draw attention to the
role of dialectic in psychological phenomena. Moreover, our reason for
drawing attention to this dialectical aspect in a book on metaphysics is
as follows: these dialectics are not, as followers of traditional schools of
thought would be inclined to think, logical in nature. They are temporal.
Fundamentally, they are successions. A function cannot be permanent;
it has to be succeeded by a period of non-functioning since energy
diminishes as soon as it is expended. As regards the phenomena of
life, it is therefore always in terms of succession that contradictions in
behaviour must be defined.

Indeed, so great is the heterogeneity of its terms that succession is
in effect discontinuity. Bergson often tones down this heterogeneity so
that as a result, succession seems like a change where things fade and
merge into one another. Thus, Bergson takes psychological intuition to
be a priori a continuous thread, imposing an essential unity on experience as though experience could never be contradictory or dramatic.   24
'A mind that simply and solely followed the thread of experience...
would see facts succeeding facts, states succeeding states, and things
succeeding things'. It seems to go without saying that things remain
beneath facts and states remain beneath becoming. Yet how can we fail
to see the isolation of essences, fixed in a way in terms of the formula
for their dimensions! Even in the most homogeneous order of thought,
you cannot go from one essence to another by continuous thought.
More generally, how can we fail to see that all differentiation in appearance and pace is the sign of absolute discontinuities, with the result that
the discontinuity of an appearance is immediately the appearance of a
discontinuity.

But Bergson goes further in his intuition of overall homogeneity. As
we have said when outlining the arguments for Bergsonian continuity,
he accepts that there is a continuous movement of exchange between
the two distinct poles of subject and object, the absence of the one being

here again automatically the presence of the other. We would only cease to think about ourselves in order to think about things, and in exactly the same way, leaving things would inevitably mean going back into ourselves. This indeed presupposes that thought is a permanent being and a temporal substance. A more functional, more phenomenist point of view would refuse to hide the very clear duality of introversion and objective thought. Where functions are concerned, in the exchange of functions discontinuity is the first datum. We shall demonstrate in very many ways that the addition of the idea of continuity to that of succession is gratuitous and without proof, always and everywhere beyond the realm of both physical and psychological experience. If we agree to examine continuity only when it is observed, we see that it only comes about in a way that is factive, belated, and recurrent. It is simply a paralysis of action that gives this allegedly primary impression of continuity. Finely detailed experience and the intuition of disorder in the mind will, though, restore us to the rhythm of *yes's* and *no's*, to life that is tried out, that is ephemeral, refused, and taken up again. In   25
other words, through different transpositions we rediscover, spread out over time, the fundamental dialectic of being and nothingness. We can thus give full meaning, both ontological and temporal, to the following Bergsonian formula: time is hesitation.

VII

Can the temporal continuum be saved by defining time as an a priori form? This method amounts in a way to substantialising time from below, in its vacuity, contrary to Bergson's method which, with duration, substantialises time from above, in its fullness.

It is quite easy to see that directly formal intuition is a total impossibility. Indeed, the prediction of the flow of time is based on the lessons of memory, its a priori only appearing a posteriori, as a logical necessity. The a priori was in fact established by Kant in a logical demonstration. It is an analytical result that will always suffer from an unresolved question: how does the synthesis of event and form come about, how does a compact element become apparent in this diaphanous environment?

We believe then that we must give ourselves rather more than just temporal possibility characterised as an a priori form. We need to give ourselves the *temporal alternative* that can be analysed by these two

observations: either in this instant, nothing is happening or else in this instant, something is happening. Time is thus continuous as possibility, as nothingness. It is discontinuous as being. In other words, we start from temporal duality, not from unity. We base this duality on function rather than on being. When Bergson tells us that dialectic is but   26
the relaxation of intuition, we reply that this relaxation is necessary to the renewal of intuition and that, from the standpoint of meditation, intuition and relaxation give us proof of the fundamental temporal alternative.

We are well aware that this dialectical function is especially vulnerable when expressed like this, and that it will make it all the easier to restate Bergson's criticism. Indeed the objection will be raised that in this form, it seems perfectly obvious that as Bergson argued nothingness is but the negation of a human expectation: when we say that nothing is happening, we are obviously saying that nothing is happening in an order of things that are more or less subjectively defined. Here then, Bergson's argument is put forward once again. Yet we shall always meet this objection with the same reply: in the order of functions, *nothing* is not *another thing*. When we say nothing in reply to a disagreeable letter it really is of little importance that we think something. In a kingdom, however many clerks are kept up on duty at night, you cannot prevent government from being suspended because the master is asleep or from always being a tissue of authority and anarchy; you can then say equally well, according to whether you are criticising or praising or to whether you are or are not a Bergsonian from a social point of view, either that a monarchy is a dispersed form of government or that a monarchy is an authority which is always ready to make itself manifest. But it always has to be recognised that continuity is a continuity which is postulated, that it takes refuge in the potential, and that it is heterogeneous to what manifests it.

This reply will not of course be accepted, and there will be the wish to materialise time and to slip into the intervals measuring our moments of weakness *things* that are given the task of having duration; we shall be drawn on to the realm of abhorred space; matter will be shown to us, placid, motionless, inert, always waiting, matter that *exists* comfortably settled in tranquil immortality. And continuous Bergsonism will move   27
imperceptibly and inevitably to an unforeseen consequence: matter is said to fill time even more surely than it does space. Surreptitiously, the phrase *to have duration in time* has been replaced by the phrase *to*

remain in space, and it is the crude intuition of fullness that gives the vague impression of plenitude. Such is the price we must pay for the continuity established between objective and subjective knowledge.

Were you to relive precise objectivation—which is the only way in which order, succession, and duration can be assessed in their relation to a reality—you would see from the very first moment that this objectivation unfolds in the discontinuity of dialectics, with all the fits and starts of contrary experiences and reflections. Between security and precision there is a dialectical relation that might appropriately be called the psychological uncertainty relation: do you wish to be sure of finding an object, in an objectivation that is certain, by attributing to it absolute and enduring existence which is entirely independent of your own duration? Are you condemning yourself to a crude definition of this object, as a whole and as the symbol of one single function? In that case, you can no doubt say that your hat is certainly on its peg, that it remains there, and that is waiting for when you go out. If it had been accidentally moved, you would at least find it in your cupboard; no essential disorder can ruin its existence or interrupt its duration. But do you wish to go deep into detail and establish precision, not with regard now to pragmatic knowledge of a particular object but rather to the scientific knowledge of a subtle matter? Here, you are obliged to imagine experiments, to provoke relations, to dynamise the multiple world of atoms. Matter crumbles away as you act precisely on it and so ends up giving only ambiguous answers to your enquiries. Its precise existence becomes as singular as your own individual existence. The coincidences of subject and object will be atomised. They will not have duration. You will no longer find subtle, precise matter always there at the disposal of experiment. You have to wait for it to produce its events. You are now in pure expectation and nothingness is no longer a disappointed expectation, absence is no longer a displacement. The microphenomenon is in fact only produced where coincidences knot together, and does not appear along the whole length of the thread. Outside these coincidences, there is no place for any experiment.

This vacuity in the development of microphenomena is something we propose first of all to record without any hesitation, taking it to be a fact. We shall then advance a step further: we attribute this vacuity to the facts in exactly the same way that modern physics attributes indeterminacy to the facts. And we believe that, in doing so, we are complying with metaphysical caution. Indeed, we do not feel we have the right to

impose a continuum when we always and everywhere observe discontinuity; we refuse to postulate the fullness of substance since any one of
its characteristics makes its appearance on the dotted line of diversity.
Whatever the series of events that are being studied, we observe that
these events are bordered by a time in which nothing happens. You can
add together as many series as you like but nothing proves that you
will attain the continuum of duration. It is rash to postulate this continuum, especially when one remembers the existence of mathematical
sets which, while being discontinuous, have the power of a continuum.
Discontinuous sets such as these can in many respects replace one that
is continuous. But there is no need to go into more detail here. Psychologically, everything can be explained in discontinuity. Besides this, we
do not even have the right to add up all the series, all too often adding
the known to the unknown. Our philosophical duty is rather to remain
in one particular series of events and seek out links that are as homogeneous as possible, linking mind to mind directly for example, without
the biological intervening.

Thus, from a particular standpoint and with regard to a particular   29
function, there can no longer be any doubt that dialectic and not continuity is the fundamental schema. As Rivers has said, 'the alternative
of two contrary reactions makes it essential to inhibit one of them'. In
other words, the contradictory interplay of functions is a functional
necessity. A philosophy of repose must have knowledge of these dualities. It must maintain their equilibrium and their rhythm. A particular
activity must have in it well placed lacunae and must find contradiction
that is in a way homogeneous with itself. Repose, while accepting contrary activities, must refuse a jumbled miscellany of activities. It is not
yet the moment to expand on these conclusions, however. Let us for the
time being stay with our temporal problem. Here then is how we would
summarise the results of our discussion of the relations between being
and nothingness.

If we take it in any one of its characteristics or in the sum of its characteristics, the soul does not continue to feel, nor to think, reflect, or wish.
It does not continue to be. Why go further in search of nothingness, why
go looking for it in things? It is in us, scattered all along our duration,
at every instant shattering our love, our faith, our will, our thought. Our
temporal hesitation is ontological. The positive experience of nothingness in ourselves can only help to clarify our experience of succession.
Indeed, it shows us a succession that is plainly heterogeneous, clearly

marked by occurrences of newness and surprise and by breaks, cut too by voids. It shows us a psychology of coincidence. But where then is the real psychological problem of *time*? Where must we seek temporal reality? Surely it must be in those knots that mark coincidence? And there must surely be a plurality in the laws of succession? And if there is   30 plurality in the laws of succession, how can we not draw the conclusion that there is a plurality of durations? Before coming to metaphysics of time, we must therefore examine specific durations. Let us first look at pure psychology, at psychology that is simply temporal. We shall then return to the problem of objective succession with an examination of the diversities of causality.

NOTES

1. Bachelard's title refers implicitly to two Bergsonian concepts. His striking use of the word 'relaxation' (*détente*) in fact takes up Bergson's definition of dialectic as only 'a relaxation of intuition' ('une détente de l'intuition') in the concluding section of chapter 3 of *L'Evolution créatrice* (Paris, 1907). The English translation of this book by Arthur Mitchell (London: Macmillan, 1911) offers the same version of this phrase while elsewhere preferring to translate 'détente' as 'detension'. Bachelard's use of 'nothingness' (*le néant*) refers to Bergson's discussion of nothingness in the first section of chapter 4 of *L'Evolution créatrice*. Arthur Mitchell's translation renders 'le néant' as 'the nought'.

2. Bachelard's footnote (amended): see Henri Bergson, *La Pensée et le mouvant. Essais et conférences* (Paris, 1933); Bachelard refers to the second part of Bergson's introduction.

3. Bachelard's footnote refers to a French translation of Schopenhauer, *Philosophie et science de la nature.*

4. Bachelard's footnote (amended): J. Wahl, *Vers le concret* (Paris, 1932).

5. Bachelard's footnote: cf. A. Koyré, 'Hegel à Iéna', *Revue d'histoire et de philosophie religieuses* (1935), 445. 'Contrary to a thousand years of philosophical tradition, Hegel thinks not in substantives but in verbs'.

6. Bachelard's footnote refers to a French translation of Von Hartmann, *Philosophie de l'inconscient.*

7. Bachelard is referring to W.H.R. Rivers, *Instinct and the Unconscious; a Contribution to a Biological Theory of the Psycho-neuroses* (Cambridge: Cambridge University Press, 1920); he quotes later in this chapter from the French translation of this book.

Chapter 2

The Psychology of Temporal Phenomena

I

For Pierre Janet, knowing always implies teaching.[1] Moreover, it is of little importance whether or not we communicate our knowledge since our innermost thought is itself 'a way of talking to oneself, a way of teaching oneself'.[2] Indeed whatever its subject, teaching has in the end to be a matter of suggesting a well-defined order for separate actions, by giving the assurance that well-ordered actions will be either objectively or psychologically successful. These actions that are *promised* by teaching are expected without our being over-particular about the intervals separating them, though we do insist on there being intervals, and we are careful to protect the promised actions from any disruption during an interval. This then is how we can schematise the trajectory uniting dogmatic knowledge to knowledge that is proven and clear, to knowledge that has indeed been confirmed by consciousness; it is in effect the path taken by real teaching.

In this respect, knowledge of time is not by nature in any way privileged. It cannot possibly be immediate and intuitive, for this would mean it was condemned to be poor and rudimentary. For it to be enriched, this kind of knowledge must, like all others, be *expounded*. Time must therefore be taught and it is the conditions of its teaching which form not just the detail of our experience but also the very phases of the temporal psychological phenomenon. Time is what we know about it. Thus Pierre Janet has very clearly stated that 'if we are talking

31

32

39

about knowledge of time, we have to arrive at ways in which we can protect ourselves against time and ways in which we can make use of it' (p. 19). We do not have the right to *realise* our ignorance and be too quick to base the development of the temporal phenomenon within us in an objective pattern. Our intuition of time is indeed too fleeting and too vague for us to forsake too soon the great lucidities of thought time, of taught time. Lastly, while the point of view adopted by Pierre Janet may at first seem artificial, it can on reflection be seen as the mark of great philosophical caution. It is good practice not to grant ourselves the right to talk about knowledge that would not be communicable.

Furthermore, we note that the first characteristic encountered by a psychologist who is expert in the study of temporal phenomena bears the mark of the fundamental duality of duration. Indeed, from the very beginning of his work in this field, time appears to Janet as either an *obstacle* or a *help*; you have to either protect yourself against it or use it according to whether you are in empty duration or in the realising instant. Psychologically, this is proof positive that there is dual behaviour with regard to the phenomena of time. Being alternately loses and wins in time; consciousness is either realised or is dissolved in it. It is therefore quite impossible to experience time totally in the present and to teach time in one single immediate intuition.

Nor indeed can duration be taught to us *directly* by our past when this past is taken as a single, uniform entity. If we adopt Pierre Janet's point of view, we will be quick to accept in fact that memory cannot 33 be taught unless it has a dialectical support in the present; the past can only be brought back to life by linking it firmly to an affective theme that is necessarily present. In other words, in order to gain the impression that we have had duration—an impression that is always singularly imprecise—we must, as with real events, put our memories back into an environment of either hope or disquiet, into a dialectical wave-motion. There can be no memory without this wavering of time, without this rippling of affectivity. Even in the past that we believe to be full, evocation, narrative, and shared secrets put the void of inactive times back in place again; as we remember, we constantly mix useless, ineffectual time with time that has been useful and fruitful for us. The dialectic of joys and sorrows is never as absorbing as when it accords with the dialectic of time. We then know that it is time that takes and that gives. We suddenly become aware that time will take from us again. Thus, reliving time that has disappeared means learning the disquiet of our own death. How striking and true is René Poirier's account of the sudden awareness we have of these fragments of nothingness and death encountered in our lives:

> Expectation is a pretext for us to experience the past. Expectation is of course disappointed desire, irritation and a feeling of powerlessness, yet more than this, it is the bitterness of time that is destroyed. Each one of the moments it erodes becomes a theme of regret. Between the living past and the future lies an area of dead life and nowhere are regret and the sense of the irreparable any stronger. It is in this way that time can be sensed by us. And it can be sensed even more in anguish and the thought of death. Not the anguish of some kind of suffering or some kind of desertion, but of no longer being anything, the anguish that a whole world is consequently destroyed. Who has not felt this thought, entering the soul like a sharp blade? It cuts so quickly that it does not even cause pain; but the heart perceives it deeper down and feels itself grow faint; thus whoever truly thinks death cannot do so without turning pale. This thought is brief, secret almost and sharp as the swallow's cry or as the bow Odysseus draws sounds to the suitors' ears, easing only when we have slowly grown inured to it or when we have high hopes. For while we can put up with no longer being ourselves, who can bear to be nothing at all if we have, just once, felt all the pain of this? Just as a horse confronted with the dead body of its fellow stops short in agitation, so our soul falters and refuses in face of such desolation.[3]

34

In teaching us all that time can break, meditations like these lead us to define time as a series of breaks. We can no longer really attribute uniform continuity to time when we have had such a vivid premonition of the weakening and failing of being.

On a gentler note, regret for missed opportunities brings us face to face with temporal dualities. When we wish to tell someone else our past, teach them our person, nostalgia for those durations in which we were not able to live will deeply disturb our understanding of history. We would like to have a continuum of acts and of life to narrate. But our soul has not retained a faithful memory of our age or the true measure of the length of our journey through the years; it has only retained the memory of events that have created us at the decisive instants of our past. When we confide, all events are reduced to their root in an instant. Our personal history is therefore simply the story of our disconnected actions and, as we tell it, it is with the help of reasons and not of duration that we consider ourselves to be giving it continuity.

Thus, our experience of our own past duration is based on real rational axes; without this structure, our duration would collapse. We shall be going on to show that memory does not even give us direct access to temporal order; it needs to be supported by other ordering principles.   35
We ought not to confuse the memory of our past and the memory of our duration. Through our past, what we know, using this word as Pierre Janet defined it, is at the very most what we initiated in time or what, in time, has collided with us. We retain no trace of the temporal dynamic, of the flow of time. Knowing ourselves means finding ourselves again in these scattered personal events. It is on a group of decisions we have experienced that our person rests.

Knowledge of duration to come would prompt the same observations; it is only by being passed on that it can be constituted; it can be passed on only by following Janet's modest but far-reaching method, translating our momentum into the language of actions we anticipate and of behaviour that is always more or less systematised. So the future we glimpse is simply the programme of promised actions. It is only our actions that we can really think in our personal future. It is impossible to realise a passive experience satisfactorily. If we envisage obstacles, it is always in terms of the reaction they will provoke in us; we always take future time in its positive moments. Thus, every intuition of the future is a promise of actions that takes no account of the duration of those actions; this intuition is limited to imagining the succession and order of active instants. Anticipating the future means fixing its pattern and disregarding the intervals of laziness, fatigue, and leisure; it means isolating its *centres of causalities*, and admitting because of this that psychological causality, as we shall subsequently establish in detail, proceeds in leaps and bounds, leapfrogging over useless durations.

In vain do we try to differentiate between understanding a process and living it, for in what we call *living a time* we must always distinguish between what we know and do not know, since our use of this phrase implies that we have immediate and silent knowledge of dura-   36
tion. We do not though *live* ignorance any more than we see darkness. The psychologist who takes us into his or her confidence and tells us that 'within myself, I *feel* time flowing without incident, without break' can only impart to us a sense of our two obscurities coming into contact, of the symphony of our two silences. This kind of psychologist seems

to us just like those bearers of secrets and mysteries who promise us treasure but only leave us with a lot of mumbo jumbo. No! In order to refer to an innermost experience, you have to be able to escape its vagueness; you have to multiply and vary examples. As soon as you do this, confidences are shown to be singular, the contingency of temporal experience is seen, and centres of psychic crystallisation are isolated. In the experience of fine detail, the very smallest of events is enriched:

Yet now as Fate
Approaches, and the Hours are breathing low,
The sands of Time are changed to golden grains.[4]

A very particular characteristic of innermost observation is that a judgement of value intervenes and clarifies what is simply a judgement of experience. It is impossible to know time without judging it. It is through this judgement that we constitute behaviour of different kinds and it is by studying these that we can truly develop a psychology of temporal phenomena.

II

Once the influence of active instants has been highlighted, we have a clearer understanding of the subordinate character of the consequences that may trail along, more or less, behind a decision. The durations of constituting acts can be either lengthened or shortened but these dura-   37
tions do not affect the essential character of behaviour. They are not attached to the act but are only its contingent and variable after-effects, with no quantitative objectivity. This lack of quantitative objectivity is the sign of an essential relativism. Why see it as the mark of a short-coming of human reason, as the price to be paid for using a method of intellectual scrutiny which does not suit its object? When an action is properly studied in a project that is made entirely explicit, the order of constituting acts dominates all else. The idea of *length* of *time* is sec-ondary. Cooperations can always shorten over-long times of execution. These cooperations give time a new dimension, a dimension of depth and intensity which, through well-regulated coincidences, gives effec-tiveness to instantaneous decisions. There is even an inverse relation between the psychological length of a period of time and its fullness.

The fuller time is the shorter it seems. This trite remark should be given prime importance in temporal psychology. It should be the basis of a key concept. The advantage would then be seen of referring to richness *and* density rather than to duration. It is through this concept of density that we can fully appreciate those regular, peaceful hours when our efforts have real rhythm, hours that give us the impression of normal time. It is to these regular rhythms in a life both peaceful and active that, following a rationalised dialectic, we relate the *length* of an inert period, of ill-constituted repose that is marked by disharmonies and shapeless becomings. Indeed, we only find that time has *length* when we find it *too long*.

The rhythm of action and inaction therefore seems to us inseparable from all knowledge of time. Between two useful, fruitful events the dialectic of the useless must be in play. Duration is only perceptible in its complexity. However thin it may be, it at least establishes itself in opposition to its limits. We do not have the right to take it to be a uniform and simple datum.

Yet we do not expect to win this argument straight away. All we wish to do for now is to secure one point in our argument, namely that duration is metaphysically complex and the decisive centres of time are its discontinuities. This is not undermined simply by saying that beneath apparent discontinuities there is still continuity in itself. We have in fact to remain at the level of consciousness. Here, discontinuous temporal behaviour of all kinds is seen to be the simplest while all continuous temporal behaviour is the most artificial.

If we examine the problem in this way from the standpoint of temporal behaviour, we shall see straightaway that the systematic use of time is hard to acquire and hard to teach. This explains then why we are often content with temporal knowledge that is general and confused. Indeed, Pierre Janet divides psychological behaviour into two very different groups, primary and secondary kinds of behaviour, and he shows that the psychology of temporal phenomena cannot be placed in the first group: 'I do not think that a single primary act can be found that is related to time . . . For there to be adaptation to time, there has to be something new, something added on. There exists therefore what we call secondary acts' (p. 53). Thus, any use of duration is difficult and unpredictable; it is a risk. Far from innermost duration being a property we own, it is a work we create and is always preceded by an action centred on an instant. It is this initial action that has first to adapt more or less precisely to spatial conditions. We must attach our time to things for it to be effective and real.

38

The objection will again be raised that an instantaneous action has a duration which it brings along in its wake in order for it to be completed. This though is a catagenic duration that is not interested in the fate of the initial act and that expends itself, following lower rhythms, in purely physiological or physical consequences. This catagenic duration has nothing in common with the anagenic duration that has to be maintained and nurtured.[5] It is really not an ingredient of the act; from the psychological view point which is ours here, it has no role to play; it can be eliminated. In any case, the duration that fades away like this, that dawdles and just follows, *is not a kind of behaviour*; it cannot be taught; it cannot therefore really be known.

In order to really *continue* an act initially adapted to space, you must therefore make another effort and add a second act. This is one of our main arguments and must, we consider, be emphasised. And once again we find new support in Pierre Janet's arguments. Indeed for Janet, effort is a superadded phenomenon of which only more developed beings are capable. Effort is dependent on the brain, which is as much as to say that it is dependent on the intellect. *Continuation* is not natural at the level of reflexes. It is the brain that in bringing reasons, adds a continuous sequence and places behind the causes of the first, initiating act causes for the development of this sequence. And this addition of reasons is what gives us strength to keep going. We only persevere with an action through a judgement of value, and in a secondary mode of behaviour. Janet writes that:

In both duration and the continuation of acts, there is a phenomenon of effort. It may seem strange but it must be said that acts become difficult because of the simple fact that they have duration. Performing an action for a quarter of an hour is not the same thing as doing it for half an hour . . . Time adds a difficulty. The first beings did not react to this difficulty; they halted their action, no matter the consequences . . . But an animal with a higher degree of development adds effort and perpetuates the action. We can say that the beginning of duration, the first act performed relative to duration, is the effort at continuity, the effort at continuation (p. 55).

In this way then, the clear, far-sighted will opens duration like a perspective; it places a sequence of supplementary acts behind the first impulse; it shows itself to be a synthesising power determining

39

40

an organic convergence. We obtain duration by progressively bringing more and more muscles into play. The analysis of *continuity* of
effort would lead us to repeat almost term for term Bergson's detailed
work developed with regard to the *intensity* of effort.[6] There is plurality in the development of continuity just as there is plurality in an
increasing intensity of effort. It can be seen that this intensity and this
continuity are in a way homologous, and that the arithmetic sum of
particular efforts which accumulate to give an intensity is dispersed the
length of a succession to give a duration. Of course, if we look fairly
closely at it, we shall see that a continuation of this kind is made up of
separate impulses. Any psychology of effort must accede not just to
the geometrisation of effort Bergson indicates when he reads intensity
in the volume of muscles progressively brought into play but also to
the arithmetisation of effort that counts the muscles which are progressively alerted.

We are thus gradually led to separate very clearly, from a functional
point of view, the will that initiates the act and the will that continues
it. Before the will to have duration was added, all there was to consider
was the reflex act stuck in the instant and taking its whole meaning from
some spatio-temporal coincidence. Contrary to this, thought, reflection,
a will that is clear, and a persevering character all give duration to an
ephemeral act by learning to add to it appropriate secondary acts. We
can therefore grasp duration as behaviour, as a work we create.

III

Furthermore, Pierre Janet's book devotes many pages to the psychology   41
of *beginning*. This is a very particular kind of psychology that could
provide the key to many problems. The mind is perhaps essentially a
factor of beginnings. Janet singles out first of all what we might call
majestic beginnings, those that inaugurate duration but do not in fact
belong to that which has duration. A *foundation-stone* laid by a government minister has nothing in common with the construction work the
builders do. This was not always the case. Certain religious introits are
real psychological preparations for mystical life, for the continuity of
religious emotion. Mauss has studied purification ceremonies from this
point of view. Simply from the psychological standpoint, one can never
over-emphasise the importance of this consecration of beginnings.

Janet concludes in fact that 'acts marking beginnings and ends have an important part to play and are highly significant', also pointing out that primitive tribes have no 'acts that introduce and that close' (pp. 62–63). Primitives limit themselves to explosive acts, that is to say to acts that really do not *continue*, psychologically speaking, since their consequences are at the very most physiological ones. In exactly the same way, some sufferers from neurosis lose continuation behaviour in which the effort that begins and the effort that continues need to be differentiated. As Janet goes on to say here, 'this is the chief characteristic of the epileptic act, an explosive act that nothing leads us to foresee, that the subjects themselves do not foresee, that has no beginning, and that comes to an end without us knowing why'.

All well constituted duration must thus be given a clearly differentiated beginning. How can it not be seen that in these splendid and solemn beginnings, the causality of reason takes the place of what was claimed to be the causality of duration? Here the supremacy of willed time over lived time is plain. May we therefore be permitted to use a paradox in order to underline effectively the causal and temporal isolation of the initial act: what makes a train work is the station-master's whistle. Conscious life is, in exactly the same way, an activity consisting of signals. It is the activity of a master. A clear intuition is a command. 42

Let us now though look at behaviour when for instance we are carried away in an impetuous rush, when we are filled with enthusiasm or are tempted, behaviour in which the beginning of the act appears to prepare as is usual what follows the act. We shall see that this beginning is however still not homogenous with what follows it. As Janet says, 'when we perform an action, energy is expended in what we do, but there is always too much of it and the excess we put in will play a part in successive movements; this is what is called élan, that is to say an impetus, an impetuous rush' (p. 65). If we look at it in this light, élan is therefore a kind of lack of economy of effort. When we are thus carried away in an impetuous rush, we think we can grasp hold of a ready-made duration; in reality, we fail to command duration, and to constitute a duration. Paradoxically, this élan brings passivity to action. We can be sure of this, for whoever is thus carried away will go astray. When later on we come to depict rhythmic life, closely bound to the temporal dialectic of repose and action, we shall see that this élan or impetus is temporal behaviour that is too simple, too ingenuous, precisely because such behaviour removes the possibility of resumption, the freedom of beginnings, and the active, polymorphous grouping of realising instants.

Let us now summarise our view of this theory of beginnings. Pierre Janet has indeed discovered a particular kind of temporal behaviour that is of the very greatest importance. If we are to teach its full significance and really master it, we have to isolate a beginning and regard it as a pure event. In other words, we need the concept of the instantaneous in order to understand the psychology of beginning. Moreover, many very different kinds of behaviour regarding beginning are only elucidated by reference to the psychology of beginning. Thus, we can only have some knowledge of an élan if we take it back to its first impulse. The conclusion must in any case be drawn that behaviour initiating duration is not simple, since we can extract a few decisive events from it which, in many respects, deserve to be described as primordial.

43

IV

We can perhaps also indirectly clarify behaviour relative to beginning by linking it to the psychology of change. Beginning and changing are far from being equivalent. A beginning can clearly be taught; a change can merely be suggested. In fact, psychologists do not yet have good knowledge of fundamental behaviour relative to change. Janet's frankness here is most instructive as it proves that our knowledge of temporal psychology is very poor. He concludes his third lesson like this: 'change is the starting point for all our knowledge of time. There must therefore be behaviour that relates to change. We do not have knowledge of this'. Janet refuses to follow Guyau and Fouillée when they refer to a *sensation of change*. 'Sensation', he objects, '. . . is a static state . . . on the table we have some red and beside it some green; we have two sensations, one red and one green. If we pass from the first to the second, we have other sensations but we only have one or another sensation' (p. 95). Here once again, it is impossible to fill a void within alterity. True methodological caution means postulating a discontinuity as soon as we are sure that a change has taken place. In fact on this occasion, the usual tendency is on the contrary to postulate an underlying continuum. Because changes lack synchronism, we think that in different areas we can find intermediate elements which make change fade and blur. Sometimes these added elements are, so to speak,

44

factors of fuzziness. We have thus overlaid melancholy on autumn so that slowly and imperceptibly, as they die, leaves can change from green to gold. We mix theatrical genres so that acting techniques can change. Yet in fact, transitions always transcend the areas that are to be linked. The soul places the confusion of its feelings beneath the mind's discontinuous determinations. We cannot therefore over-emphasise the importance of Janet's observation that 'change is nearly always consonant with feelings, and very often with the feeling of sadness. Change is in fact quite sad; in all its forms, it is nearly always a dying'. Thus, we base all the events in our lives on the continuum of our sorrows; we translate into the emotional language of continuity what would be more accurately expressed in the clear and trenchant narrative of objective events. Continuity is but our emotion, our unease, our melancholy, and the role of emotion is perhaps only to blunt ever-hostile newness. From the standpoint of temporal behaviour, we can thus conclude along with Pierre Janet that 'feeling is a regulation of action' (p. 99).

V

It is not only change that allows us to accede to discontinuous behaviour. Clearer psychological cases can be found that sanction the teaching of a real *behaviour relative to nothingness*. Pierre Janet has in fact placed much emphasis on *deferred behaviour*, on the interruptions of an action whose consequences are postponed to the future. Now, deferring an action means that its causality is suspended, and that continuous duration has its chief function removed. No longer does the stream impel the stream. We are free to decide what is urgent.

This is not isolated behaviour; it interferes with kinds of behaviour that at first sight seem distant from it. Thus in Janet's theory, memory is influenced by deferred behaviour. Janet argues quite rightly that memory is a faculty that develops late, that is indirect and linked to reason, and in step with social organisation: 'Bergson usually accepts that an isolated person has memory. I do not share this view. A person who is alone has no memory and does not need it' (p. 218). He goes on to say that 'the act of memory is an act that is relatively rare . . . I cannot argue that we have a universal memory, that in this memory we embrace all that we have seen. This is entirely imaginary; here the metaphysical principle has filled pure memory, and it is a totally

45

arbitrary supposition' (p. 255). We shall see memory constituted in a truly reflective duration, in a recurrent time. Indeed, memory does seem to become clearer through choices made, and to gain strength through its framework, not through its matter. Memory uses the temporal enjambement of deferred action. In other words, *we remember an action much better by linking it to what follows it rather than to*   46 *what precedes it.* This paradoxical conclusion has to be drawn if we accept that all clear thought—thought that is therefore taught—has to be based on behaviour. Now behaviour of different kinds is only possible if it gives itself a future and makes its finalism explicit. Lived duration does reveal the matter of memory to us but not its framework, and it does not allow us to date and order memories. Yet an undated memory is not a real memory. Far from being pure memory, it remains a reverie tinged with illusions. It is in fact because we are able to create a void before our action—in other words, because we can defer it, and to use still another phrase, because we can shatter its catagenic causality—that we have the means of giving our memories a framework. We keep coming back to the very significant idea of the social frameworks of memory developed by Halbwachs in an impressive book.[7] What makes the social framework of memory, though, is not just history lessons but far more the will to a social future. All social thought is pulled towards the future. All forms of the past must, if they are to give us truly social thoughts, be translated into the language of the human future. As a result, even in individual terms, it is impossible to refer purely and simply to an innermost intuition, to knowledge which the past is seen as writing passively in our soul. Thus, Janet does not hesitate to write that 'deferred action is in my view memory's true starting-point' (p. 232).

It is in deferred action that we become clearly aware of negativism, since here negation becomes a kind of behaviour. We really do create a void before action that we defer. No doubt Bergson would say that we hasten to fill this void by performing other actions. Yet the dialectic is not so dense and the attitude of refusal can be seen organising itself as refusal.

The problem of recalling memories would also be clarified by paying more attention to the *instant* in which memories are really fixed. We   47 would then see the role played by the co-ordination of new events, the almost instantaneous rationalisation of events that are linked together in a complex memory. Before we deal with the preservation of memories, we must study their fixing since they are preserved in the actual framework in which they are fixed, as more or less rational totalities. Thus, Pierre Janet proposes quite correctly that the problem of amnesia be combined with that of *amnemosynia*, in other words that more importance be attached to the *absence* of memory than to *loss* of memory. We would then understand the role played by dramatic thought in fixing our memories. We retain only what has been dramatised by language; any other judgement is fleeting.[8] If there is no spoken, expressed, or dramatised fixing, a memory cannot be related to its framework. Reflection must construct time around an event at the very moment when the event takes place so that we can rediscover this event in the memory of time that has disappeared. Without reason, memory is incomplete and ineffectual.

A study of the temporal conditions of the fixing of memories would also show us the memorising power of an event that is expected and desired. It seems that expectation creates a void in us, that it prepares the resumption of being, and helps us to understand destiny; in short, expectation makes temporal frameworks in order to receive memories. When the event takes place that we have clearly expected—paradoxically again—it appears to us as clearly new. Nothing happens as we had expected it to; thus the event both satisfies and disappoints our expectation, justifying the continuity of an empty rational framework and imposing the discontinuity of empirical memories. All who are adept at savouring expectation even when it is anxious will acknowledge its skill in creating the picturesque, the poetic, and the dramatic. Out of the foreseen, it makes the unforeseen. Oh the heady pleasures of meetings we await! We only need love just enough, fear just everything, and wait in quaking anxiety and the person who is late will suddenly seem more attractive, more certain, and more loving. As it bores into time, expectation deepens love. It places the most constant love in the dialectic of instants and intervals. It restores to faithful love the charm of newness. The events we anxiously awaited are thus fixed in memory; they take on meaning in our lives. The clearest memories are thus the ending of a day's drama, of an hour's drama. They are the reward for our earlier refusal to live anything other than what we desire. It is by deferring mediocre actions and by being utterly intent on foreseeing the unforeseeable that we prepare ourselves for being thoroughly contradicted by happiness. In contradicting us, the event is fixed in our being. Dialectical assimilation is the very basis for the fixing of memories.

48

There can be no emotional memory without an initial drama, without surprise by opposites.

Although our argument that a framework is preliminary to memory has been intentionally developed first of all in the domain of affectivity which least favours our point of view, it will be clearer in the domain of purely intellectual memory. All acts of memory are inseparable from a schematisation that, by dating events, isolates them. It empties them of their duration in order to give them a precise place. This schematisation is like a rational canvas, like an outline for the narration of our past. This outline is thought to link facts; in reality, it separates them. When for instance narration shows two events to be in a logical sequence, it demonstrates that the second is produced from the first by deferred behaviour. In exactly the same way, if we are really to understand the duration that lies open before us, we must live the promises of the future through thought; in the place of the very vague and meagre impression of lived experience we must set our decision to follow a life-plan. Our feeling of duration is in proportion to the number of projects we have. True possessions, those we believe to be substantial, are those we can carry forward to the future. This carrying forward cannot be done according to a schema of homogeneous continuity, since everything that secures it has a reason producing it. I am perfectly happy to say that tomorrow I shall enjoy such and such if reason proves that tomorrow my enjoyment will be the greater. The organisation of memory is parallel to this organisation of present duration. Conditions of recall are the same as the conditions that construct fixing. It is an intolerable misuse of analysis that makes us separate the fixing and the recalling of memories. Memories are fixed only if they first comply with the conditions of recall. We therefore only remember by making choices, by filtering out the lees of life, by cutting facts from the current of life and putting reasons there instead. Facts remain in memory thanks to intellectual axes. This insight of Janet's is very profound: 'what has created humanity is narration, and not by any means recitation' (p. 261). That is to say, we do not remember simply by repeating but we have to *compose* our past. Character is a biased story of the self. Moreover, Janet makes it very clear to us that the work of memory does not finish with the act of memory:

> it is not over when the event ends, because memory is perfected in silence. A young child tries out the romance he or she is getting ready to tell their mother . . . This is the gradual perfecting of memories that takes place

49

little by little. This is why a memory is better after a few days than it was at the beginning, it is better made, better wrought. It is a literary construction that is made slowly and is gradually perfected (p. 266).

Events do not therefore settle along the length of duration like direct and natural gains. They need to be ordered in an artificial system—a rational or social system—that gives them meaning and a date. This is why delirium that is not sufficiently systematic leaves no trace at all. Janet in fact observes that 'after even complex epileptic delirium, there is no memory. This is not because it is complicated but because sufferers have not constructed the act of memory, being too stupid to do so in their delirium' (p. 224). 50

Thus, memory is a work that is often difficult, and is not a datum. It is not a possession we have at our disposal. We can only realise it by starting from a present intention. No image arises without a reason, without an association of ideas. A more complete psychology ought, it seems, to emphasise the rational or occasional conditions of a return to the past. Psychoanalysis in particular would benefit from stressing the *present* importance of past traumas. As Janet might say, when we think we are telling our dream by *repeating* it we are in fact *narrating* it. This is not far from being a justification, a proof. To psychoanalysis we could therefore add another aspect. It asks why the patient had this dream. To this we should add: why does the patient tell the dream? This would bring us back to the study of the present conditions of psychosis.

Indeed for Janet:

the problem of recall is above all a problem of initiating something and stimulating it. Why indeed should the person who has deferred an act cease to defer it? . . . The merit and miracle of memory is to have constructed an act that is initiated by something that is not precise, that has not yet happened. It is preparation for obeying a signal other than the familiar ones.

It is a mechanism waiting to be set in motion by a future coincidence. Memory is therefore not realised all on its own, through an inner surge. It must be differentiated from reverie precisely because true memory possesses a temporal substructure that reverie lacks. The images of reverie are gratuitous. They are not pure memory because they are incomplete, undated memories. There is no date and no duration where there is 51 no construction; there is no date without dialectic, without differences. Duration is a complex of multiple ordering actions which support each other. If we say we are living in a single, homogenous domain we shall see that time can no longer move on. At the very most, it just hops about. In fact, duration always needs alterity for it to appear continuous. Thus, it appears to be continuous through its heterogeneity, and in a domain which is always other than that in which we think we are observing it. Always and everywhere, the phenomena of time appear first of all in a discontinuous progress. They reveal to us an order of succession. Nothing more and nothing less. In particular, the way they are linked is never immediate. In many respects, succession is free; it accepts suspensions of action, it accepts clear heterogeneity, as we shall see when we look closely at the problem of causality in relation to time.

NOTES

1. Marie-Pierre-Félix Janet (1859–1947), on whom Bachelard draws extensively in this chapter, was at the time a leading authority in France in the study of mental illness, neurosis, psychological medicine, and the development of intelligence. After studying at the École Normale Supérieure in Paris, along with Bergson, Durkheim, and Jaurès, he directed the neuro-pathological service at Le Havre hospital while teaching at the local *lycée*; he went on to study medicine, taking his doctorate in medicine in 1893. With Charcot, he directed a laboratory in experimental and comparative psychology at La Salpêtrière hospital in Paris; he taught at medical schools in London, Harvard, Mexico City, Rio de Janeiro, and Buenos Aires, and received honorary degrees from sixty-three universities. He was publicly honoured in France, becoming president of the Académie des sciences morales et politiques in 1925 and Commander of the Légion d'honneur in 1927.

2. Bachelard's footnote (amended): P. Janet, *L'Evolution de la mémoire et de la notion de temps* (Paris: Chahine A, 1928), p. 22. All Bachelard's quotations from Janet in this chapter are from this book; he sometimes omits page references.

3. Bachelard's footnote: R. Poirier, *Essai sur quelques remarques des notions d'espace et de temps*, p. 64.

4. Bachelard's footnote refers to the French translation of E. A. Poe's *Politian*. I quote from the original text of Poe's play here (scene 6, lines 39–41).

5. A reminder of the etymology of catagenic and anagenic may help to clarify Bachelard's meaning: their prefixes are formed from the Greek words for 'down' and 'up' respectively, 'genic' being derived from the Greek verb 'to

be born', 'to become'. Bergson in fact uses both words in *L'Evolution créatrice* (chapter 1, section 2) when discussing the development of living things, taking these terms from a contemporary American biologist, Cope.

6. See in particular Bergson's *Essai sur les données immédiates de la conscience* (Paris, 1889), chapter 1.

7. Maurice Halbwachs (1877–1945) published in the field of sociology; Bachelard would appear to refer here to his book *Les Cadres sociaux de la mémoire* (Paris, 1925). Influenced by Durkheim and Bergson, Halbwachs first taught philosophy at Caen University and then sociology at Strasbourg University; he took up a chair in sociology at the Sorbonne in 1935, and a chair in 'collective psychology' at the Collège de France in 1944. He was arrested by the Gestapo in 1944 and deported to Buchenwald, where he died the following year.

8. Bachelard's footnote: as Jerusalem says in *Urtheilsfunction*, 'language always dramatises the simplest judgements' (p. 9).

Chapter 3

Duration and Physical Causality

I

In actual fact, all causality is displayed in the discontinuity of states. We 52 show one phenomenon to be a cause and another to be an effect when we draw a line round each of them which defines and isolates them, giving each the unity of a name, and revealing the essential organic character of each. If we refer to a clearly defined effect, then what we are wanting to do is rid ourselves of the accidental. If we refer to a cause as being certain, then our wish is to establish a hierarchy of appearances in the phenomenon. A Bergsonian will no doubt see this static double designation as simply proving that linguistic and spatialising needs dominate our intellect. He will call on innermost intuition to follow the causal continuity from one phenomenon to another. Yet this continuous and wholly inner link will, in its turn, only be expressed by a general word, and without objective proof We shall never succeed in *unrolling* the sequence of causality. As soon as we analyse a cause of this sequence, as soon as we specify its development, we are dividing up this cause of a sequence into successive states; and in affirming that these states are linked, we are oddly enough eliminating the duration linking them. We have made the cause into so complete a phenomenon that it seems the cause must take place all by itself and bring about the effect in a time which may be short or long, but whose length we have little interest in determining.

We trust people will not be too quick to accuse us of abstraction! We trust that, in particular, this will not be seen as surreptitious support for 53

Bergson's thesis of a mathematical time that can only represent the flux of phenomena by a series of cross-sections! No, cause and effect are neither of them simply temporal breaks. They both have a particular temporal structure. For each of them, this structure constitutes duration. What we are affirming though is that this duration, which is in some way immobilised so as to constitute cause and effect separately, cannot in any way succeed in linking the effect to the cause. We do not have to take account of the duration in the cause or of the duration in the effect in order to link them temporally. Within the cause, duration is simply preparation. Beyond the effect, duration is simply a damping down, abatement. A phenomenon that was prepared at great length does not react any more strongly than one whose preparation was rushed. Physical causality is not quantified by duration. We must always reach the point at which we establish the phenomenon cause and the phenomenon effect as two separate states, and since their particular duration is ineffective we should, as it were, empty them temporally. We are moving towards the rationalisation of causality. We very gradually come to regard the cause as a principle and the effect as a consequence. The link between them is thus as contemporaneous as it is deferred. Rationalised cause and effect are fixed in their individuality. From the moment we deduce one from the other, we get rid of the irrationality of their temporal link; this link is only contingent, only an initiating act. We nearly always have means at our disposal with which we can accelerate the effect when we have really understood a cause. If we prepare sugar in granulated form for the lecturer, we are giving him or her, as an initiating act, the means to drink a glass of sugared water without having to wait for it to dissolve.[1] There is nothing really objective in time other than the order of succession. In any case, coming back to the firm ground of real proof, in the realm of objectivity that is discussed and experiment that is demonstrated, phenomena are presented as successive and discontinuous. A historical account of physical phenomena is full of interregnums that scientists quite correctly ignore; they can be ignored and therefore they must be ignored.

54

II

We shall now go on to see that the verification of causality comes in an atmosphere of negations, in a kind of logical void, which again stresses the isolation of cause and effect.

Let us establish this proof with an example that is as simple as possible, and where the positive aspect is at first sight particularly clear and obvious. Kant takes the following judgement as an example of a close synthesis: the sun warms this stone. However, this positive form conceals a countless number of negative judgements. Indeed, the judgement of experience is not simply a posteriori; it comes late. It closes a polemic. And it is indeed what is absolute in negation that gives the principle of causality here its necessary character: we are only sure of what we negate. Let us try to follow once again here the polemic of refusal that prepares our support for causality.

Generally speaking, before all else the application of the principle of causality comes down to negating the activity of substance. Far from the category of substance being, as Schopenhauer argued, a replica of the category of causality, the category of causality fulfils its function by negating the causal action of substance. A phenomenon is the cause of *another* phenomenon. Things pass the cause to each other; they do not give rise to it. A cause of itself is either a tautology or else a god. It is perhaps in this way that *causality* and *participation* will most clearly be seen as contradictory. In so far as a quality is thought of as participating in the activity of substance, it eludes causal analysis.

In addition, the affirmation of the action of something *alien* is still not entirely positive or at least, it is only positive in so far as it is imprecise and general. As soon as this affirmation becomes precise, it brings negations into play. A phenomenon's features can only be distinguished by differentiating. When we postulate the effectiveness of a cause, we are noting the ineffectiveness of various supposed causes. Thus, saying that the sun warms this stone means proving firstly that it does not warm up all by itself, due to the activity of substance, and then secondly that it is not warmed by another source of heat.

Furthermore, our argument would be more pertinent if it could be developed with reference to a more scientific example since this would then give us a better sense of the indispensable polemical role played by false hypotheses. There is however a methodological interest in approaching the problem through an example that is as familiar as the one chosen by Kant. Indeed, familiarity increases the falsely positive appearance of our experience. In our dealings with the slow, drab world of crude experience, we soon unlearn how to be surprised. We come to think symbolically because the phenomena manifesting the whole have the immobility of symbols. We rely on sensory wholes because we imagine that these wholes are syntheses. It is in this spirit

55

that the following objection to our ideas will again be raised: is there not the synthesis of the phenomena of light and the phenomena of heat when the selfsame ray touches both our hands and our eyes? Or again, expressed in more realist language, is it not obvious that the wave's vibration *is* at the same time both heat and light? Now, this sensory combination sets us on the path towards identity and in so doing invites us to intellectual inertia. By eliminating differences, the declaration of identity terminates this experience. Yet who does not see that such an experience is far from being even half-begun? The reply is so clear   56
though that it seems definitive. It is made so quickly that it seems immediate.

On the contrary, active reflection ought to lead us to draw the conclusion that a synthesis based on experience cannot be an immediate datum. This experimental synthesis is not simply a posteriori from the rational point of view, because of the gratuitousness of experience. It is also a posteriori because of the intervention of polemical reason. At the very root of heuristics there lies a whole eristic, and there is a whole dialectic of the true and the false at the origin of our judgements of experience. An attempt at synthesis always lays the foundations for its success by opposing previous failures. The cause cannot, in essence, be the subject of an intuition since, given that the idea of an effect should be more complex than the idea of a cause, the differential of newness evident between cause and effect should be the subject of discursive thought, of thought that is essentially dialectical. Intuition can doubtless shed light after the event, and it then has the force of rational habit; it cannot however illumine the initial quest. Before intuition, there is surprise.

Thus, eliminating errors reveals the cause. It is in this elimination, which has been made a very conscious process, that the real pedagogy of causality lies. In order fully to understand the cause of a phenomenon, it is indeed in our interest first of all to refuse explicitly the various causes that might occur to us. In reality, in the history of our learning, there has never been an immediate phenomenon that could be attributed to a precise cause. A precise cause is always a hidden cause. And this observation will appear all the more important as we grow more aware that the search for causes always has a reaction on the task of description. In discerning a cause, we are distinguishing between the characteristic features of the phenomenon we are studying. Every efficient cause becomes a reason for explaining a structure. We often only understand the structure through the cause. It is often the propagation of physical agents that draws the lines of matter. Thus, structure is as 57 much efficient cause as formal cause. There is therefore a kind of correspondence between form and development. A geometric hierarchy is in command of a temporal order of succession. And vice versa, causal discipline requires a spatial order. A complete phenomenology is one that is both formal and causal.

The regularity of phenomena therefore entails a logical preparation of experiment. A causal law can only proceed safely in so far as it is protected from disturbance. There can be no detection without protection. In order to follow the logical isolation of cause and effect, we need only reflect on any law of physics. We shall see that thought that is entirely verbal, that is condensed into the identity of a commonplace phrase, will be segmented into two distinct images whenever there is the slightest attempt at precision. And this segmentation will appear to be the two phases of a process which has a before and an after. For example, if I declare from the outset that a falling stone is attracted by the earth, I have the impression of a unified phenomenon. However, in this dogmatic response, intuitive thought is not really active. As soon as I wish to make my thought precise, I shall be drawn along a discursive path and it will not take long for me to see the time of explanation polarise and gather around two distinct centres. Along with my thought about the real action of the earth on the moving body there will in fact be at the same time thought about potential action, action entirely preliminary to real action. I shall analyse reality—what everyday language calls reality—through possibility. I shall then bring in the static idea of the gravitational field. I shall understand the influence of the earth in its possibility rather than in real causal development. In particular, it is by going more deeply into this idea of field, an idea that is wholly intermediate, that I shall gain better understanding of the detailed phenomenon of gravity. It will also give me a better grasp of this phenomenon's conditions of differentiation, for example sensitivity to a change in gravity with altitude and also the correct definition of the vertical, a definition in which I shall give the centre 58 of the earth a role to play. This gives us a pretty good idea of how the cause is fleshed out, of how it is organised and completed. Only when I have studied the field in this way, determining the conditions and the limits of its uniformity, will I introduce the stone into this particular field. And through the cooperation of the moving body's mass, the field will become a force. The synthesis giving the effect will then be seen as having in a way one more dimension than the cause. The cause will only act when something is added, and this will be to the benefit of a convergence of conditions. The realisation of the cause in order to give its effect is therefore an *emergence*, a composition. Thought that is subtle and detailed, proved and taught will lead us to establish the heterogeneity of cause and effect. The better we teach, the more we shall differentiate. The pull of gravity will be analysed in two phases[2] by relating two objects, the moving body and the earth, And also by distinguishing between the time of the possible and the time of the real. And the possible opens up a discursive enquiry in which polemical reason is given full scope. The study of the potential mathematical functions which are the basis of the mathematical physics of fields is, like it or not, founded on the metaphysical idea of the potential. We rediscover the ancient way of thinking that is seen in the passage from the potential to the actual, with initially the metaphysical heterogeneity of the potential and the actual, cause and effect. It is perhaps by going more deeply into a theory of causality of this kind that *minimum emergence* could be found, to be precise that which appears in time as the first action of time, as a slight accentuation of reality that produces a definitive effect.

III

So far, we have only discussed the problem of causality with regard to its application or indeed, to put it even more simply, with regard to its explanation and the way it is presented to us. In the end, what we have done is indicate how causal relations are *taught*; we have not determined what these causal relations are in themselves. There is no doubt in our view that teaching conditions are eminently conditions of objective thought. This however is not the place to develop this personal theory and we know readers have long been nursing an objection: what does it matter how causality is proved, since beyond the discontinuity of proofs there will always be the continuum of the real cause that operates in the double continuity of space and time. It is this key objection that we must now confront.

Let us first of all observe that when we think causal development in a continuum which we cannot exhaust, we are seeing mystery in this

59

development and exaggerating the richness of becoming in exactly the same way that naive realism exaggerates the richness of substance. In other words, we ascribe too much action to time when we make it the support and the substance of action. If temporal action really *formed* the phenomenon, we would not understand the resistance that forms put up to deformation. In fact, causality and form unite and dominate time and space. René Poirier put this very well when he wrote that 'time and space are thus imbued with causality; causality is a natural part of them and transfigures them' (*Essai sur quelques remarques des notions d'espace et de temps*, p. 17). Indeed, causality in its many forms brings many reasons for relations, links, and successions, and by doing so makes time and space organic. In this way too we see how causality imparts knowledge of varied time to us. This is of course by no means the conclusion drawn by Poirier. His analysis leads him instead 'to restore the time and space in which things are to their task of being impassive spectators, and to despair of becoming and of our understanding of becoming'. Yet this despair does not afflict the agent of scientific change, the scientist who by associating the different forms of causality comes to construct from nothing phenomena which are precise and predicted. Contemporary science has time as well as space as a variable; it can make time effective or ineffective with regard to qualities that have been differentiated. When we have a better understanding of how to produce frequencies, we shall gradually be able to fill time discontinuously in the same way that atomism has filled space.

From a certain point of view, a technique that produces becoming should be able to suspend the action of time. In order to have the same effect, there must be the same cause. In order to have the same cause, time must not act on the well-defined phenomenon; we must be able to restore the cause in its identity in order for the effect to be restored in its identity. Now, the permanence of the cause can only be clearly and certainly realised if we start from rationalised phenomena. We can only completely define that which we understand. Only the truly organic cause is able to give a well-defined effect. The principle of causality is always seen as playing between two distinct and very clear-cut figures, eliminating both accidents and details.

In other words, there is a hierarchy in becoming just as there is a hierarchy in the essence of being. The more purely a cause realises its essential scientific schema, the more regularly will it determine its effect. The experiments that work best in physics are not the simplest ones but the most organic. They are those in which experimental precautions have been taken *systematically* and detail has been confined to its role as detail, and where we are sure of the *non-causal character of detail.* When the polemic of precaution has been carefully conducted, we feel safe from accidents; we feel able to initiate behavior relative to scientific beginnings and to postpone the rationalised phenomenon to a time we have determined. We have only to compare the continuous 61 waves used in radio to the always irregular, always accidental sparks eighteenth-century electric machines produced in order to understand what makes for a phenomenon that is mastered by time. The modern system seems in a way like a temporally closed system, one which is represented in its rhythms just as a thing is represented in its spatial limits.

Having thus made a kind of relative assessment of the temporal effectiveness of a phenomenon's diverse causes, we are now entitled to reconstruct complex becoming without basing ourselves on an absolute time which is external to the system and which claims to be valid for all parts of the system. Each part of a system has its own appropriate temporal rhythm specific to the variables as they change and develop. If we do not see this, it is because we most often conduct an experiment from one particular point of view and affect only one particular variable. And we think everything else has been left as it was. Temporal correlations are evident however in very many cases and they prepare a pluralist theory of time.

On other occasions, we go to the opposite extreme and introduce continuity of development in order to link two different states. This continuity of development should make clear to us the heterogeneity of durations affecting different features of the phenomenon. In actual fact, we postulate continuity between two slowly modified aspects of a phenomenon because from other points of view, it is not hard to see rapid modifications. These rapid modifications act as a transition; they are examples of transitive states. However, heterogeneous development is not a real link. It is very instructive to see that development is the price we have to pay for a complexity we have not analysed. Thus, we only need to *complicate* a kaleidoscope by adding many smaller fragments to the larger ones for it to seem to change and develop continuously. The jerkiness of events will then dissolve and fade away because they 62 are so numerous.

In what way then can an experiment dealing with fine detail be helped or clarified by the postulate of temporal continuity? A duration that nothing analyses can always be accused of only being valid as 'duration in itself'. It will not be the phenomenon's duration. Microphenomenology ought not to attempt to go beyond describing the order of succession or more simply still, beyond enumerating possible cases. This enumeration will then require a time that is purely and simply statistical, having no more causal effectiveness. This brings us to one of the oddest fundamental principles of contemporary science: statistically, the different states of a single atom in duration and a group of atoms taken at a particular instant are exactly the same. If we reflect on this principle, we ought to be persuaded that in micro-physics, antecedent duration does not *propel* the present and that the past does not weigh upon the future. Since there is complete geometrical homology between an individual's *development* and the *state* of a society, conditions of structure can be exchanged with conditions of development. In other words, here once again causality is efficient causality just as much as it is formal causality. There is another conclusion to be drawn, for according to this principle the atom's becoming is obviously applied on to a number and not on to a continuum; the atom's becoming just hops about since this becoming finds its equivalent in a countable plurality of atoms in different states, and since the successive states of an atom are found when we go from one atom to another. The temporal dialectic is therefore simply the development of the ontological dialectic.

IV

Furthermore, there is a break between an experiment taking an overall view and one that goes into fine detail and this break radically alters the conditions of objectivity. Let us clarify this alteration. When I say 63 that a phenomenon taken as a whole *changes* from state A to state B, what I mean is that between A and B there are myriad details and accidents which I ignore but which it is always in my power to indicate. If, though, I study fine structures at the limits of experimental accuracy, I must take a new postulate into account: *the detail of detail has no experimental meaning*. The detail of detail falls in fact into the absolute nothingness of systematic error, of the error imposed by the requirements of detection. It is at this point that the dialectic of detection comes into play, with the rhythm of all or nothing. Discontinuous number replaces continuous measurement. There is no longer anything continuous apart from error; error is simply a halo of possibilities surrounding measurement. And as for determinations, these are quantified. We can thus understand that when we take causality in the forms in which it is experienced in fine detail, it is fragmented. Indeterminacy is an almost immediate consequence of the quantum character of measurement. There is nothing that permits us to put a temporal continuity into place in order to analyse discontinuous transitions. If we do so, we are taking duration *from the outside*, regarding it as a useful function, a synthesis that is more or less arbitrarily imposed on the dispersion of phenomena. Duration certainly cannot be read in a real analysis of phenomena.

There is even a kind of contradiction in postulating the inexhaustible diversity of a phenomenon along with the rigorous identity of detection. We have in fact attained a level of knowledge at which scientific objects are what we make them, neither more nor less. We have mastery over objectivity. The history of the laboratory phenomenon is very precisely that of its measurement. The phenomenon is contemporaneous with its measurement. Causality is in a way solidified by our instruments. Objectivity becomes the purer as it ceases to be passive in order to become more markedly active, as it ceases too to be continuous in order to become more clearly discontinuous. We *realise* our theoretical thought by degrees. We end by extracting complex phenomena from their own particular time—a time that is always vague and indistinct—in order to analyse them in an artificial time, a time we determine, the time of our instruments. We are able to slow down, accelerate, or immobilise the most varied temporal phenomena. With stroboscopy, we can detach and select particular instants in a rhythmic phenomenon. We can produce an accurate history of these elements that have been isolated from their context, by linking them to elements taken from outside the whole contexture of reality. The continuity we make in this way has very plainly no connection with a real continuity; it has however all the attributes of a real continuity. Philosophers should reflect on the ease with which the time of instruments can thus be substituted for the time of phenomena. This ease of correspondence between the 'real' phenomenon and the instrumental phenomenon of stroboscopy should suggest the idea that the essential function of duration is doubtless purely and

64

simply 'correspondence'. When we make two orders correspond, we are giving them the same order of succession. Once the correspondence has been completed, duration is no longer of much use. This is why the temporal homologies outlined by stroboscopy are accurate and conclusive. They shatter duration. They retain causality however. If lastly we observe that from some points of view our senses are more or less well-adjusted stroboscopic instruments, then knowledge of duration can more easily be ascribed to a construction. Our everyday knowledge of temporal phenomena is produced by an unconscious, lazy kind of stroboscopy. Duration is the stroboscopic aspect of a general change; it is the separation of fluid elements from stable ones. If we believe in the permanence of things, it means that we always open our eyes at the same phase of their rhythm.

Thus, a detailed study of causal relations teaches us to exercise choice in the succession of phenomena. Our action on the temporal characteristics of a phenomenon is far more effective than it would first seem. If we have the ability to associate the spatial and temporal characteristics of a phenomenon, then through material intermediaries we can, as it were, frame temporal phenomena. Rhythm is imprisoned in sound boxes. When we see a rhythm preserved in a radio aerial, we cannot stop the image of a reciprocal action between the geometric and the temporal from intruding into our thought. It is therefore in our best interests to regard things as truly the products of stationary waves. Periods are spatio-temporal functions. They are the temporal face of material things. As it vibrates, a thing reveals both a temporal and a material structure.

If we now add to this that periods are immediately translated into the language of frequencies and that frequencies appear as relative to one another, we see the absoluteness and the continuity of time not just fade but disappear. In any case, the continuity of absolute time that might serve as a basis for differentiating between periods is no longer the immediate continuity yielded by crude observation. The causality studied on the basis of frequencies is indeed in play *above* the continuity that is *postulated* as fundamental to the duration of a period. In particular, the study of this causality through periods and frequencies could, in our view, be limited to statistics of periodic events. The *regularity* of an isolated vibration is postulated perfectly gratuitously, for in fact it is only the *frequency* of groups of vibrations that is used. Moreover, it must be noted that most phenomena which are explained by frequency are explained by a fairly large number of frequencies. The slow periods of astronomy do not serve here as an explanatory motive. If the earth is considered as it moves in its orbit, it does not 'vibrate'. It follows its path. The time of astronomy is therefore not yet 'structured'. If we consider the monotony of planetary motion, we can well understand how a uniform, continuous time came to be ascribed to it. It is precisely a time in which nothing happens. It is an inadequate schema for establishing the realism of rhythm. 66

When we go deeply into the finely detailed forms of multiple causality, we become aware of the value of temporal organisations. We are less and less tempted to regard causes as just breaks in a general Becoming. These causes constitute wholes. They act as wholes, spanning useless intervals, without regard for images representing time as a flux whose entire force lies at its limits. Causal energy is not located on a causal wave front. The cause requires organic conventions. It has a temporal structure, a rhythmic action. It belongs to a spatio-temporal structure.

Alongside the cause's organic character and in connection with this character, we must also make way for the kaleidoscopic and discontinuous character of material change. Causal relations can thus gain in clarity when we study them from the arithmetical standpoint. There must be an advantage in *arithematising causality*. In this respect, the new quantum science is preparing for us special ways of studying that must sooner or later come together in an arithmetic of effective instants.

NOTES

1. Bachelard's implicit reference here is to Bergson's use of the example of a sugar lump in *L'Evolution créatrice* (Paris, 1907) in support of his argument that time is 'lived': the time I wait for the sugar to dissolve is 'my impatience', 'my duration'; it is not mathematical time (*L'Evolution créatrice*, chapter 1). Bachelard argues against Bergson's interpretation in *L'Intuition de l'instant* (Paris, 1932), chapter 1; his argument is presented and discussed in M. McAllester Jones, *Gaston Bachelard. Subversive Humanist. Texts and Readings* (US: University of Wisconsin Press, 1991), pp. 29–35. It should be noted that in Bachelard's phrase 'sugar in granulated form' there is an implicit cultural reference: in France in his day, lump sugar was more usual and he thus differentiates

himself clearly from Bergson by using this phrase. In *L'Intuition de l'instant*, Bachelard had in fact referred to Bergson's sugar ('le sucre') as a lump of sugar ('un morceau de sucre').

2. Bachelard's phrase here 'en deux temps' is an example of his humorous—and untranslatable—word-play since it refers to a two-stroke internal combustion engine, while also enabling him to maintain his focus on the discussion of time.

Chapter 4

Duration and Intellectual Causality

I

Our intention in taking the problem of temporal effectiveness into the   67
domain of physics was simply to confront possible objections and to
conform to philosophical custom: it is in fact usual to want time to be,
to begin with, an objective power and to see movement as giving us the
clearest measure of duration. It seemed to us that in this domain itself,
temporal links were neither as strong nor as uniform or general as they
are said to be. The thread of time has knots all along it. And the easy
continuity of trajectories has been totally ruined by microphysics. Reality does not stop flickering around our abstract reference points. Time
with its small quanta twinkles and sparks.

It is not though by reflecting on physical phenomena that we can
really feel the metaphysical duality of duration. In objects, in fact,
breaks remain accidents and elude all attempts to systematise them.
In higher psychic activity, breaks are on the contrary inseparable from
reasons; or better, the small variations in energy involved in higher
psychic activity bring about new ideas. Here we can say that small   68
variations have great effects. Our mind, in its pure activity, is an
ultra-sensitive time detector. It is very good at detecting the discontinuities of time. For this to happen, all we need do is turn aside from
all practical chores and all social cares, and listen to time's cascades
within us.

Furthermore, physical or physiological phenomena would always teach us to submit to time and be an object among objects. A whole aspect of the phenomenology of time is obscured when we limit ourselves to reflecting on the development of phenomena. Their kinematics can so easily be described that we come to believe that their dynamic character is less certain, less general, and more concealed. In actual fact, the history of science shows fairly clearly that dynamics is added to kinematics as secondary and derived knowledge, which is harder and more fallacious.

Yet if we leave objective reflection and come to our own innermost experience, everything changes and what was obscure becomes clear; the experience of inner dynamics now moves to the fore while that of our movements seems derived and secondary. From this standpoint, movements seem to us to be simply the consequences of our decisions, taking into consideration—which is very important—the *difficulties* of carrying out our decisions. We must not neglect this very first and wholly intellectual aspect of the difficulty of our acts. It is this aspect that can best teach us about active time. In any case, when the dynamic and the kinematic are studied in ourselves they should give two very different impressions of time.

And there is more besides. In us, the dynamic initially appears in the form of impulses, jerks, and rushes of feeling, in short in a discontinuous form. And to illustrate the dialectic of the continuous and the discontinuous in relation to time, the simplest is perhaps to confront our movements and the original order of the will that governs them. The dualism of the continuous and the discontinuous is then homologous with the dualism of things and the mind. Having already argued in a previous chapter that continuous effort is behaviour that is secondary, learned and difficult, we need only look now at the impulse in its dynamic aspect as an active element. Yet if continuous movement is a physiological consequence and if the essential element of an act is the impulse, is it not in the organisation of impulses that the control of intelligent action must be sought? We should therefore establish, as Paul Valéry has so well put it, *an algebra of acts*. An action thus appears as having a necessarily complex formula with many articulations, and having between the impulses well-defined dynamic relations. Intensity then has a primary meaning and no longer just one that is derived, as in Bergsonian theories. Quantification happens at the level of the will and no longer at that of the muscles. In

69

this indirect way, the intellect acquires real causality. It is the intellect that turns away contradictory actions and determines active convergences. This intellectual causality must doubtless take account of both physical and physiological causality, but there is even so a place for psychological rationalisation that will give particular effectiveness to the intelligent act.

II

It is by analysing the complex of strength and skill that in our view a first assessment can most easily be made of this clearly determined effectiveness, which is already visible at the level of the will. A skilful psyche is one that has been educated. It manages energy. It does not allow it either to flow away or explode. It proceeds by making small, very separate movements. With consciousness of skill there comes an entire geometry that is necessarily composed of straight lines and edges, in contradiction to the sweet unconsciousness of grace.[1] Grace must not be wished; it has lines, not axes. It is pure quality; it   70 condemns quantity. It does its best to erase the discontinuities of the learning process and gives unity to the most varied actions. Skill must on the contrary keep the fundamental hierarchy of many movements. It is kaleidoscopic. It is strictly quantitative. Grace has the right to make mistakes; for grace, error is often a whim, embroidery, a variation. Skill must not amuse itself. And why should skill seek to combine together the decisions composing it? It indeed runs a risk in moving away from the pure arithmetic of separate wills. From its standpoint, these curves with their lazy inflections are lines of lesser thought, of lesser mental life. They appear when there is a subsiding, when the conscious being is returning to reverie, allowing itself to be overcome and vanquished by external resistances. These curves could doubtless be regarded as more *natural*; but this is precisely the proof that they require less consciousness, less supervision, and less mental input. For skill, nature both within us and outside us, is first and foremost an obstacle. It is the innermost obstacle especially that makes skill a real energetic controversy, a real dialectic.

Rignano has very perceptively indicated this fundamental dualism in the perfecting of certain skilful movements. Let us for instance look again, as he does, at the skills involved in a game of billiards; we shall see this psychologist concerned not with peripheral descriptions of effort, but instead with describing the *central* structure, just at the level of the dialectic of more and less:

> The billiard player who has already aimed his cue at the ball is impelled above all by the desire to make his stroke and he prepares himself for this. But too much tension in his arm muscles leads him to fear making too strong a stroke as he has already done a short time before this; then, thanks to this antagonistic activity, his muscles relax a little. But the lessening of tension that the player then feels taking place which, in its turn, is linked to the memory of some previous stroke he bungled by not giving the billiard-ball a sufficient turn of speed, awakens the contrary fear in him of using too weak a force. The to and fro movements as his arm brings the tip of the cue either closer to or farther from the ball before the stroke is made will be seen by someone watching the game as reflecting the very swift succession of conflicting feelings produced one after the other. These feelings grow weaker or stronger by turns in order to reach the end result of imparting the required force to the ball.[2]

71

Rignano has only studied the quantitative framework of muscular energy here; but he has clearly shown that the intelligent use of strength needs *two* contrary reference points, more and less. He has also clearly shown that the impression carried to the centre for an over-tense muscle determines, through reflection, a relaxation that is exactly the opposite of the action prepared by physiological causality. Physiological causality ought not to wait; it ought to initiate the stroke that is too strong. Yet reflection imposes an interval of inaction and then an opposite conclusion. The action takes place through a contradiction. Skilful will is never good will; in order to act, skilful will has to go through bad will. Skill can really not be conceived as something unitary, taking place in an unbroken duration. We do not really have at our disposal a substantial, positive, and unified memory that would allow us to reproduce exactly a skilful action. We must first weigh up contradictory memories and achieve the balance of opposite impulses. These discursive operations make time uneven; they break the continuity of natural development. There can be no real certainty in the success of a skilful action without consciousness of errors that have been eliminated. Then thought time takes precedence over lived time and the dialectic of reasons for hesitating is transformed into a temporal dialectic.

72

III

If we do not always see the importance of the role played by the hesitation imposed by reflection at the threshold of actions, this is because we rarely study the psychology of actions which have been well learned and well understood, and which are fully conscious of their success. Usually in fact, we endeavour above all to link the psychology of intelligent behaviour to that of behaviour that is more or less instinctive, more or less natural. This is no doubt a useful thing to do. Yet in making it the only thing psychology does, we may be led to disregard the specific meaning of certain problems. In particular, artificial action, action that is marked by reflection, is often action with no stimulus or even against the stimulus, or simply when there is a stimulus. It therefore introduces a whole range of stimulating events in which the most diverse causalities interfere. We thus begin to see how a whole psychology of mental liberation could be worked out by disentangling all these interferences. In order to study the first stage of this liberation from the stimulus, we could look again at everything Rignano has to say about senses that act without making any contact, far from the insistent hostility of the world of objects. We would see that these senses 'most often give rise to the particular state of an affective tendency that is initiated and held in suspense' (p. 45). Here we have a kind of false equilibrium uniting opposites and permitting almost instantaneous effectiveness to be given to a decision that has been well prepared, but left waiting to be implemented. Starting from this stage, which is still entirely physiological, we realise that what initiates action is not just the effectuation of physiological coincidences. There has to be permission to act, and the mind must lend its full support to being. We only feel this support, we only feel the mind's presence, in the repose that precedes action, when the possible and the real are clearly compared. The mind's support is therefore strictly contemporaneous with an impulse or better, with a kind of impulse, the impulse of an absolute beginning. Consequently, while in its elementary form behaviour relative to beginning was still dependent on objective signs, in a purely intellectual form the will to begin appears in all its gratuitousness, and is fully conscious of its supremacy over the mechanisms that have been set in motion. The physiological causes of the sequence of actions cannot therefore be confused with the psychological causes of its initiation. A philosophy

73

that erases this duality in causes is based on a dangerous metaphysics, on a unity that is not sufficiently discussed.

Were our criticism accepted, we would suggest that a schema of initiating acts should double every motor schema. The psychology of a composite action could not in fact be taught unless the order and dynamic importance of decisive instants had first been fixed. The action will then be executed more or less swiftly. Order thus dominates duration. Order really gives us the algebra of action: the figure follows from it. An *analysis situs* of active instants can disregard the length of the intervals just as the *analysis situs* of geometrical elements disregards their magnitude.[3] The only thing that counts is the way they are grouped. There is thus the causality of order, the causality of the group. We are all the more aware of the effectiveness of this causality as we move higher up towards actions that are more composite, more intelligent, and that we keep under close surveillance.

When a motor schema is seen to be dominated by its schema of initiating acts, it is soon nothing more than an unconscious system. Its functioning can be slowed down or impeded by tiredness, wear and tear, and illness, and Bergson has proved with great clarity that these kinds of destruction do not in any way entail the ruin of pure memories. Our conception of a rationalised memory, made more alert by the elimination of all memories of duration so that only the memory of the order of elements is preserved, would lead us to conclude that pure memories remain valid not just in themselves but in their grouping. The schema of initiating acts would allow us to become aware of the preservation of composite memories, of functional memories. We can also understand that a schema of initiating acts is able to transfer its power from one mind to another. Through this schema, we suggest, we keep close watch and we command. The importance of this action of interpsychology must not be underestimated since this aspect is reflected in every human being, and an inner dialectic of command and execution makes us see very clearly within ourselves the supremacy of willed time over lived time.

74

IV

It is in fact when we become aware of the order of initiating acts that we achieve self-mastery in a complex, difficult action. When we entrust ourselves in this way to the supremacy of intellectual over physiological causality, we are protecting ourselves from indecision and mastering the hesitation to which every detail of the act could give rise. The whole is in command of the parts. Rational coherence gives cohesion to development. For instance, a long speech will be held together by the rational coherence of its well-ordered reference points. Should there be a moment when language is unclear, or when an obscure detail or an anacoluthon[4] in expression occurs, the confusion will only be short-lived and will not destroy the continuity of the whole. The plan of this speech acts as a unifying principle, as a formal cause. It is a schema of initiating acts. It is held in the mind through a whole which is made up of brief, simple signs.

Moreover, this oratorical schema provides an excellent illustration of the causality of order. We also know that a whole speech can be distorted when the order of two arguments is reversed, even when these arguments are very independent of one another. In the same way, we realise when we think about it that what makes the best links is not a continuity that advances by degrees, that is contemporaneous with real and more or less contingent development. Looking for this gradual continuity would mean putting ourselves on the level of an inattentive, unintelligent audience that cannot really appreciate intellectual continuity. No, good links are those established between arguments which are clearly distinguished and carefully categorised, in accordance with the admirable principle of dialectical rationalism so well expressed in Jacques Maritain's maxim: 'Distinguish in order to unite'.

When action, thought, and speech are gathered in this way on their successive summits, they therefore take on a continuity of composition that very obviously commands the subordinate continuity of execution. Yet this continuity is even more appreciable and appears even more effective when we are no longer content to present it as a gradation that is entirely logical and static. It has indeed a dynamic quality. It brings with it *rapidity*. This is a point of view we too often fail to study. Experimental psychology no doubt takes many measurements of reaction times, but these always concern reflex acts or simple acts. It does not turn its attention to the duration in which rather more complex problems are resolved. This duration of composition seems in fact to have no objective meaning; a thousand incidents may occur which slow it down, and in particular intervals of leisure or nonchalance between the composing acts seem to go on *ad libitum*. In short, the *continuity*

75

of composition remains logical and we do not think of bringing out its psychic value as we ought to do when we consider the psyche as clearly engaged in an effort to attain maximum consciousness. And yet if we 76 are willing to reflect on ourselves, we shall soon be aware of a very particular characteristic given by the *rapidity* of discursive thought as it links the stages in a well-made argument. This rapidity is not just speed. It also has the characteristics of ease, euphoria, and momentum, characteristics which could give a very precise meaning to a truly specific kind of energy that might well be called rational energy. This dynamism of understanding requires consciousness of the possession of a form. We are not aware of it at a first attempt and do not at first see its value. Intellectual causality must indeed be set in place. This dynamism is contemporaneous with a new beginning. It is therefore structure and construction. It is a cause that has the ability to start all over again after it has had its effect. It is a rhythm. We master it by preparing the succession of intellectual events, attaining in this way what is truly succession in itself, totally emptied of the durations of development and expression and relieved, as far as it is possible to be, of the burden of all physiological obligations.

All psychological durations, clearly represented as they are in carefully thought out convictions, are constituted in this way, thanks to the heterogeneity of form and content and to a rational law that experience endlessly confirms. Durations are first of all *formed*. They are fleshed out and filled later. What is in them is not always what really constitutes them. At the very most, the apparently continuous duration of the subordinate psyche, the monotonous and formless psyche, consolidates the more broken form of intelligent thoughts and actions, broken by all its lacunae. Yet willed order remains quite plainly the antecedent temporal reality. When we ignore this essential distinction, we lack the hierarchical principle we need in order to analyse temporal knowledge correctly. We do not see that the story of a journey is a function of its geography. It is not possible to describe something properly if we do not already 77 possess a pre-existing principle for finding reference points. Nor is it possible to describe temporal psychology if we do not give decisive instants their major causality.

A theory of filling of this kind does not moreover mark a return to a metaphysics of fullness, since there is always the heterogeneity of container and contained and also the supremacy of the form. The fundamental nature of this duality will perhaps be better understood from examples of temporal consolidation in which the heterogeneity of container and contained is especially clear. In dealing with this problem, we shall draw on a theory of consolidation developed by Dupréel in work that is of great significance. This theory provides us with good examples of the active constitution of duration. It shows very clearly that duration is not a datum but something that is made. We shall devote a chapter to this theory in order to preserve its unity.

NOTES

1. With this reference to grace, Bachelard reminds readers of Bergson's view of grace, developed in the first chapter of his *Essai sur les données immédiates de la conscience* (Paris, 1889), implicitly arguing against him.

2. Bachelard's footnote: Rignano, *La Psychologie du raisonnement*, p. 51.

3. *Analysis situs*, meaning 'analysis of place', was the traditional name for what is now known as topology, the geometry of place which studies the qualitative and the relative positional properties of geometric objects, irrespective of their form and magnitude. The homology to which Bachelard refers in this book is an aspect of topology.

4. An anacoluthon is defined as a lack of grammatical sequence.

Chapter 5

Temporal Consolidation

I

Dupréel's argument starts from the same point as ours, from the opposition of instants and intervals. In other words, he distinguishes between the time we refuse and the time we use, between on the one hand time which is ineffective, scattered in a cloud of disparate instants and on the other, time which is cohered, organised, and consolidated into duration. Dupréel rightly takes it to be a fact both fundamental and patent that when we describe the psyche temporally, we have to postulate lacunae. We can subsequently study how these lacunae are filled, and we can claim that they were made to be filled; obviously though, a void must be postulated between the successive states characterising the psyche's development, even if this void may be simply a synonym for the difference between states that are differentiated. There is moreover a metaphysical reason that backs up this methodological need for intervals: directly or indirectly, we must give way to finality, that is to say to the determination of the present by a future which is by no means close and to which we ascribe above all a certain depth. If we are ready to accept the existence of a hierarchy of active instants we shall come, quite naturally, to accept the fundamental reality of a temporal framework. The adaptation of subordinate psychic events to this framework will thus be a recurrent adaptation. This kind of adaptation with its orderly sequence and strict hierarchy will not be subject to the objections raised to an adaptation that is continuous and obscure, in which there is nothing to

78

79

81

indicate the importance played by instants that are really active. It is akin to that adaptation through the formal cause which is fundamental to Bergson's theory of creative evolution. And it is this recurrent adaptation that Dupréel so aptly describes as *consolidation*, studying it in a most thought-provoking book.[1] Anyone reflecting on Dupréel's method will soon be convinced by the clarity that familiar examples bring. We ourselves have been encouraged by our reading of his works to go on with our own apparently perilous method, which amounts to explaining the lower by the higher and lived time by thought time. If Dupréel takes certain social forms to be 'the biological in its nascent state', then we may be correct in carrying out a similar reversal with reference to the psychology of duration and in affirming that thought time is lived time in its nascent state; in other words, thought is always in some respects a trying-out of or a first move towards a new life, an attempt to live differently, to live more or even, as Simmel has argued, a will to go beyond life. If we think time, it means that we place life in a framework; it does not mean we take a particular aspect of life that will be better understood the longer we have lived. This almost always means intending to live differently, to rectify life first and enrich it later. Thus criticism is knowledge, and criticism is reality. We shall see these two moments of a meditation on time clearly apparent as we follow Dupréel's philosophy of time which is so simple yet at the same time so profound.

II

If we are to understand Dupréel's book properly, the best thing is to start with the image he has suggested in order to define 'that which coexistence consolidates' since the latter will also be very helpful in enabling us to grasp the reality of 'that which succession consolidates' in which we are especially interested:

> Generally speaking, whenever something is made there are two very clear successive states: first of all, the parts of the object to be constructed are assembled and placed in the order in which they should remain. At this point though, this order is only maintained by external and provisional means. Only in a second, definitive state will the parts themselves, through an internal adjustment, keep the position in relation to each other that is in the finished object. When for instance a crate is to be made, for

a few moments it is the maker's hands which hold the pieces of wood against each other that he is going to nail. Once these have been hammered in, the crate holds together all by itself: it has gone from the first to the second of the two states to whose succession we have just referred. This is clearer still in the moulding process; the duality of time in this process is marked by the duality of the mould and the object that is moulded. Before the cement is poured in, the object's parts are already placed in the correct order, but the force maintaining this order is external to them, it is the solidity of the mould (p. 11).

We thus pass from an ephemeral order to one that lasts, from an entirely external and contingent order to one that is internal and necessary. Dupréel therefore puts forward his argument concerning *that which succession consolidates*:

> Is it not the case that what happens with regard to spatial relations also happens with regard to temporal ones? Are not certain orders of succession first secured by an external cause and then later come to hold together, that is to say come to reproduce themselves, through an interaction of conditions which is less foreign to them, through a cause which has in some way become internal? (p. 16).

81

This very pertinent question immediately suggests to us the possibility of a theory of the gradual interiorisation of life and thought. In our view, this kind of inside made from the outside—the very opposite of what happens when a substance expands—is particularly suited to giving us the schema of a duration that is enriched with events and that constitutes distinct temporal realities.

Let us therefore see the constitution of all that succession consolidates, of all these *objects* of the psychology of duration; let us see how duration is moulded in precise temporal forms. Here again, the best thing is to start with the very simple, very clear example that Dupréel gives:

> Industry in the true sense of the term, the activity of humans who are associated and directed by aims, furnishes us at once with examples of *things that succession consolidates*. A clock is simply and solely one of these. By the time the person who made it thinks about setting it correctly it is already something that coexistence consolidates and that in addition has now to be made into something succession consolidates. For the finger of

the clock to move round the dial twice every day, no more and no less, the clock-maker must speed up or else slow down the clock's ticking in accordance with a chronometer that is itself adjusted to the earth's rotation. Here, the external order of support is the earth, the chronometer, and the clock-maker all combined. Once the movement has been duly perfected, the order to which it corresponds has become internal to the mechanism; the operation that transfers and fixes has been completed and an order of succession has been *consolidated*.

This order has indeed been brought from the outside, going from the whole to the part.

We can now find this process of temporal consolidation every time an order stabilises, be this in society, memory, or reason. Thus Dupréel   82
shows us that the transition from a social custom to a truly moral dictate takes place through consolidation: 'the external order of interests has been replaced by the internal order of conscience'. Interiorisation is again seen very clearly here. When we move on to the psychology of individuals, interiorisation may well be harder to discern, but if Dupréel's schema is kept in mind we shall, even so, be able to see it at work. For example:

When children learn a fable by heart, they first find the order of its lines on the pages of their reader. Every time memory fails them, they glance at the text; they read and bit by bit every lacuna disappears from their memory. The order of the printed word is banished. *Knowing* means that we have learned; the order of what we know was first upheld by a force external to our understanding, the latter having consolidated this order for itself, thus making any external pattern superfluous (p. 19).

We can see very plainly here that order is not purely and simply registered but reconstructed with a faithfulness which has been thought out and willed, and sustained by reasons for coherence which are specific to the learner. Were we to take examples where the mind is more free, we would see that consolidation takes place on more subjective hierarchical bases.

A whole theory of knowledge could easily be developed by emphasising the process of consolidation. We would see in particular that, as Dupréel points out in a note, induction is a consolidation of experience and deduction a consolidation of induction. This general application would, we believe, also lead to a conclusion we wish to indicate: all the

means by which we consolidate, however *artificial* they may appear, are in the end entirely *natural*. They seem artificial to us because we still   83
see in them the mark of our own effort; we indeed feel that the given comes to us in temporal and spatial disconnectedness, or at least that its original solidity is shattered by the very slightest precise use; we therefore come to consolidate the given; we consolidate it in our own way, using mnemonic methods and rational ones equally. We readily accuse these attempts to consolidate of deforming nature. In making this kind of criticism, we do not see that nature always needs to be *formed* and that it seeks forms through human activity, in fact. If, as we ought, we put human activity back into nature's line of action, we shall acknowledge that intellect is a natural principle and that what is formed by reason is quite obviously formed by a force of nature.

We can therefore affirm that consolidation applies naturally in the realm of knowledge as in the realms of life and social activity. This consolidation really does govern the constitution of forms. To be very exact, it is the sum of formal and material causality. We shall understand it still better when we have reflected on this highly significant corollary stated by Dupréel: *'there is only growth through intercalation'*. It is impossible to attach too much importance to this principle which in our view casts an unexpected light on all theories of evolution. All that grows is first enriched within itself. It is this inner enrichment that determines growth. Growth is but a consequence. As Dupréel says very perceptively:

> Life has not moved from an original nucleus towards an indeterminate development; it seems to have resulted from an advance from the external   84
> to the internal, from a state of dispersal to a final state of continuity. It has never been like a *beginning* from which a *consequence* results, but it was from the first like a framework that is filled, or like an order that has gained in consistency through, if we may be permitted to use the expression, a kind of gradual stuffing . . . Life is certainly *growth*, but all growth that is *in extension*, like fabric that stretches or individuals that proliferate, is only a particular case; life is essentially growth through *density*, an *intensive progress* (pp. 38–39).

One might be tempted to see this intensive progress as a substantialisation of intensity, but we must be very clear that in Dupréel's theory there is nothing at all mysterious about it. Indeed, this intensity is analysed from a standpoint that is clearly formal and, so to speak, geo-metrical. Its development is presented in a totally discursive way, in both its detail and its rectification.

If we thus consider a temporal movement in its analytical aspect, it will not at first be correct to describe it as continuous; or at least if the continuity of a temporal movement is to be truly reliable, truly real and sure, then the intervals must be properly organised. Without this internal organisation, the form will not hold; it will disappear like some failed first attempt at something. Continuity must therefore always be upheld by solidity. We shall then come to discover variety in continuity itself, just as there is variety in the process of consolidation. For example, we shall give continuity to a temporal movement either by increasing the density of intercalary acts or else by regulating when these intercalary acts appear. Basically, rich duration and regular duration are two very different types of continuity. If our argument is correct, disturbances of temporal psychology may be of two principal kinds according to whether the frameworks of temporal consolidation are affected or, on the other hand, to whether the internal organisation of intervals is disturbed. There will thus be two kinds of bradypsychy[2] according to whether the cells remain empty or are shattered by a disordered organisation.

In any case, it seems to us that this metaphysics of consolidation and   85
intercalation legitimates and completes our fundamental intuition that all progress takes place in two times: the posing of a form and material intercalation are the two inevitable moments of all coherent or rather cohered activity, of all activity that is not purely and simply made of accidents. Only this kind of cohered activity can be renewed and can constitute a precise temporal reality.

III

In addition to this attempt to describe the constitution of something succession consolidates, that is to say the determination of a real temporal *object*, Dupréel's philosophy also examines the precise nature of the fabric of time. Here, Dupréel develops a critique of causality and reveals that it is, of necessity, full of lacunae. He then shows the intervention of probability in the lacunae of the causal sequence. He thus prepares a renewal of probabilism to which we wish to draw attention.

The foundations of this new probabilism will be found in two works, a book and an article.[3] Dupréel in fact argues that between cause and effect there is always a necessary distinction; even when this distinction would simply be the result of the need to put forward two definitions to determine the two phenomena under consideration, it would nonetheless still establish the existence of a logical distance. To this logical distance there always corresponds an interval of time. And this interval, even where causality is concerned, is fundamentally different in essence from causality. Indeed, it is in the interval of time that impediments, obstacles, and deviations *can* intervene and these will sometimes shatter causal chains. This possibility of intervention must be wholly regarded as a pure possibility and not as a reality we do not know. It is not because we do not know what will intervene that we fail to predict the absolute effectiveness of a given cause; rather, it is because there is between cause and effect an entirely probable intervention of events which are not in any way at all connected to the causal datum. In particular, we shall never have the right to *give ourselves* the interval. In science, we can construct certain phenomena, we can protect the interval from certain disturbances, but we cannot get rid of every intervention of unforeseen phenomena in the interval between cause and effect.

We are well aware, up to this point, of the close connection between Dupréel's ideas and those of Cournot.[4] But in Dupréel there is an added nuance which is decisive. Here, what determines chance is not, as in Cournot, the *accidental* crossing of two causal lines which each have rigorous continuity. Indeed, chance conceived according to Cournot's intuition would not in any way be open to probability information; it would be pure accident. Dupréel's theory sheds light which helps us to understand that the probable already stems from any causal chain taken in isolation:

> Cournot's way of speaking is too dependent on traditional language and still gives the impression that chance or the fortuitous *is itself only an accident*, and that as the exception to the rule there are sequences of facts that are possible without it intervening, and that are complete without it. The fortuitous fact is for him constituted by two elements of another nature, by facts that are caused and by their encounter. This is the preconception we must avoid; the fortuitous is not a parasite of causality, it is by rights in the very texture of reality . . .

86

87

In fact, all known reality is known in the form of a series of successive or concomitant events, that are perceived as regular terms of a same order and between which there is an interval that is always taken up by events of some kind or other. If events are only considered at the ends of the ordinal series, we in no way reach *a reality* but only an abstract schema, since it is bad metaphysics to postulate *an ad hoc bridge*, as causality in itself would be, which would join the ends of the series closely together by missing out the interval of time or space that is always there between them. If on the contrary we say we can reach and define the pure interval, that is to say a kind of reality beyond any ordinal series in which it is framed or with which it contrasts, then this would mean we were chasing shadows: we cannot grasp hold of the indeterminate as such (*La Cause et l'intervalle*, p. 23).

Thus, Dupréel has no trouble proving that his argument takes proper account of *all* reality, that is to say of, at one and the same time, cause and obstacle, fact and possibility, what happens and what could happen. If we only stress the necessity of causes and mentally get rid of accidents that really do hamper the development of this necessity, then we are indeed going in for scholasticism and realising an abstraction. If we take a cause that is as effective as we could wish, there will always in the development of its effectiveness be free scope for possibilities of halts or deviations. We must take account of these possibilities where they are encountered, in the forms in which they are encountered, and in the intervals in which they intervene so as to modify the expected effect statistically. Even more important, we must take account of them in the description of reasoned behaviour in which possibilities become the elements of decision.

Lastly, to take a new concept of Dupréel's, this possibility which is taken in the causal sequence without going outside the causal chain appears as a very simple, very pure probability, as *ordinal probability*. A purely ordinal probability is intrinsically marked by the simple interaction of plus and minus signs. The event it indicates just seems more probable than the opposite event. It is not quantified. The quantification leading to the calculus of probability only appears when we can enumerate possible cases, for example in the case of the most schematised phenomena such as those produced by combinations of inter-actions. Where phenomena separated by a great logical distance are concerned, as in the phenomena of life and of the psyche, the question arises as to whether it will ever be possible to calculate them.

In fact, it is *ordinal probability* that determines the processes of an individual psyche.

With just this ordinal probability, we have the link that will enable us to understand temporal sequences in higher and higher 'emergence'. Indeed, whenever an emergence, a phenomenon that goes beyond its given, makes its appearance, we can see that evolution is ever more clearly determined by probability and not by causality alone. In other words, we see that the living being and the thinking being are less involved in necessities than in probabilities. And this involvement safeguards freedoms precisely because it is only ordinal probability that is concerned here. Quantified probabilities, accounting for results in retrospect, can be expressed in the form of laws which are apparently necessary. Before a decision is taken, ordinal probability faces the alternative presented by behaviour that is still to be initiated: it inclines, but does not necessitate.

Once probability in this very simple form of ordinal probability has been re-established in behaviour, then as Dupréel has in fact said, considerations of finality are no longer to be banished from ideas about life. Even when the end may not be clearly seen, ordinal probability is clarified to some degree by the end we glimpse. The end has an ordinal probability that is stronger than just any chance whatsoever, and a stronger ordinal probability is already an end. The two concepts *end* and *ordinal probability* are closer to one another than *cause* and *quantified probability* are. With this new concept, many contrasts between vitalism and mechanism grow blurred. Followers of Dupréel's philosophy find it gives them schemas that are sufficiently flexible for them to understand the connections at the different levels of emergence. We shall now pose the problem in a slightly different way by studying temporal superimpositions.

NOTES

1. M. Dupréel, *Théorie de la consolidation. Esquisse d'une théorie de la vie d'inspiration sociologique* (Brussels, 1931). Bachelard's quotations in sections I and II of this chapter are from this book; he omits some page references.

2. 'Bradypsychy' appears to be one of Bachelard's neologisms; 'brady' is derived from the Greek 'bradus', meaning slow, Bachelard's word being modelled on French terms such as 'bradycardie' (bradycardia, a slowing of the heart) and 'bradypepsie' (bradypepsy, a slowing of the digestion).

3. M. Dupréel, *La Cause et l'intervalle ou ordre et probabilité* (Brussels, 1933); 'La Probabilité ordinale', *Recherches philosophiques*, 4 (1934–1935). Bachelard's own article 'Idéalisme discursif' appeared in the same volume of this journal. *Recherches philosophiques* was founded in Paris in 1931 and set out to encourage new directions in philosophy, giving preferential treatment to phenomenology in the wake of Husserl's Paris lectures of 1929 and the publication of his *Méditations cartésiennes* in 1931. Bachelard contributed to volumes 1, 3, 4, and 6, and there are several references in his books to articles published in this journal: see, for example, chapter 6, notes 3, 4, and 8.

4. Bachelard refers here to the applied mathematician and philosopher of science Antoine-Augustin Cournot (1801–1877), who is best known for his work on the interpretation of the calculus of probability.

Chapter 6

Temporal Superimpositions

I

If we study the aesthetics of music and poetry from the stand-point of time, we shall come to recognise the multiplicity and truly reciprocal correlation of rhythms. In exactly the same way, if we make a purely temporal study of phenomenology we shall come to consider several groups of instants, several superimposed durations, which sustain different relationships. While it may until now have seemed to physicists that there is one single, absolute time, this is because they initially took up a particular experimental stance. With relativity has come temporal pluralism. For relativity, there are several times that doubtless correspond to one another and conserve objective orders of sequence but that do not however keep *durations* which are absolute. Duration is relative. Even so, the conception of durations in relativity theory still accepts continuity as an obvious characteristic. This conception is in fact based on intuitions of movement. The same is not true in quantum physics. Here, physicists are on a new level and what determines their intuitions is not *movement* but *change*. All the problems we encounter in assimilating quantum theory stem from the fact that we are explaining a change in quality with intuitions of a change in place. If we try to think about pure change, we see that here continuity is simply a hypothesis—and a very bad hypothesis—since a continuous change can never be experienced. It can therefore be presumed that the development of quantum physics will necessarily lead to the conception of discontinuous durations which will not have the linking properties illustrated by our intuitions of continuous trajectories. Qualitative becoming is very naturally a quantum becoming. It has to move through dialectic, going from the same to the same via the other.

Were we able to found a wave and quantum biology on the basis of wave and quantum mechanics, we would soon find ourselves confronting pulverisations of time which, in order to determine temporal effectiveness, would require special statistics relative to the microphenomena of life. Lecomte du Nouy's book has a number of interesting suggestions to make regarding this.[1] For Lecomte du Nouy, the time of physics is simply what enwraps individual biological times in the same sense that a light wave enwraps a multitude of elementary wavelets. Continuity would therefore be the result of temporal superimpositions. We could go further and say that the time of a tissue is continuous because of the statistical regularity of the necessarily irregular times of its cells.

Yet philosophers do not have to go deep into these provisionally forbidden regions in order to accept both temporal pluralism and temporal discontinuity. The difficulty they have in keeping to one particular line of thought shows them fairly clearly a time that is made of accidents and far closer to quantum *inconsistencies* than to rational *coherence* and real *consistencies*. This mental time is not, we believe, just an abstraction 92 from life's time. The time of thought is in fact so superior to the time of life that it can sometimes command life's action and life's repose. Thus, the mind's time pursues its action deep down, acting at levels different from its own level of sequence. It also of course acts at a purely mental level, as we have tried to establish in our study of intellectual causality. These small glimmers of light do not indeed suffice to clarify for us the multiplicity of our experiences of time. They can though let us glimpse an aspect of our argument: time has several dimensions; it has density. Time seems continuous only in a certain density, thanks to the superimposition of several independent times. In the same way, all unified temporal psychology is necessarily full of lacunae, necessarily dialectical. This is what we shall again try to prove in this chapter, using new arguments.

II

Were we to venture to relate our own views to a major theory, it is here that we should recall a number of Hegel's themes. Our wish having been to write as a teacher, teaching how to make a first attempt at sketching temporal waves, we did not want to start with metaphysics as difficult as Hegel's. We also feared that we would be accused of tending towards logicism and of having a dialectic that was more logical than temporal. And yet how inappropriate such an accusation is when we turn to Hegelian method! This is what Koyré has recently shown in a short article that is as good as a whole book.[2] The concrete character of Hegelian idealism has never in fact been so well and so quickly established:

> What Hegel seeks to give us . . . is by no means an analysis of the *idea* of   93
> time. Quite the opposite, it is the *idea* of time—an abstract, empty idea—
> that Hegel undertakes to destroy by showing and describing to us how
> time is constituted in the living reality of the mind. Is this the destruction
> of time? Is it its construction? Both these terms are inappropriate. It is not
> a matter of destroying, even dialectically, or of constructing. It is a matter
> of bringing out and discovering—and not postulating hypothetically—in
> and for consciousness itself, the moments, stages, and mental acts in and
> through which the concept of time is constituted both in and for the mind
> ('Hegel à Iéna', p. 444).

Koyré goes on to show the actual, active character of Hegelian dialectics. The latter are not logical terms limiting one another and offering us, as from the outside, the contradiction of their aims. It is indeed the mind that grasps itself in the two associated dialectical actions. This being so, we can understand that in trying to move up to pure mental time we reach the regions of both inner contradiction and the contraction of being and nothingness. As the soul thinks of itself, it makes itself adopt the attitude of refusal since it rejects objective kinds of thought; it therefore goes back to nothingness within itself, returning to that fundamental disquiet of the mind which Hegel has so clearly described. A further lesson of Hegelian metaphysics is that in giving ourselves being by refusing being, our restoration is assured, together with the automatic recovery of minimum repose. Lastly, the whole problem of the aggregation of dispersed and disparate mental acts is posed in Koyré's striking conclusion here. In describing 'the constitution of time or more accurately the self-constitution of the concept of time', Hegel does not envisage:

> an analysis of the idea of time, an abstract idea of an abstract time, of the
> time we find in physics, Newtonian time, Kantian time, the strictly linear

time of formulae and of watches. Something else is in question here.
Time itself is in question, along with the mental reality of time. This time
does not flow in a uniform way, nor is it a homogeneous medium through   94
which we would ourselves flow; it is neither the number of movement
nor the order of phenomena. It is enrichment, life, and victory. It is itself
mind and concept.

Here we glimpse the superimposition of concept and life, of thought
and time. Were we able to make beautiful temporal figures from our
psychic activity, if we could in other words *really consolidate* the temporal structures of the mind, there is no doubt that we would ease that
Hegelian disquiet born at the level of mental time with the awareness
of how difficult it is to remain at this level. This disquiet does not have
its roots in life, for submission to life at a lower level, to the flimsy
continuities of instincts, would at once erase it and would give us that
lower form of repose in which we cannot remain when once we have
left it. It is indeed the prerogative of thought to ease this disquiet and
grant us true repose. Our sense of a duty to seek the higher, rare and
pure rhythms of mental life is therefore reinforced.

III

We shall therefore try to explore superimposed times psychologically.
Simply because they do not have the same principles of sequence,
thought time and lived time cannot be postulated as being naturally
synchronous. There is a kind of vertical relativity that gives pluralism
to mental coincidences and that is different from the physical relativity which develops at the level where there is the passage of things. It
is very hard to define this cohesion of coincidences but a number of
psychologists do have some inkling of it. Thus, Alexandre Marc writes
that:

Pragmatists are apt to proclaim the primacy of action while in reality sub-
ordinating action to the category of the useful, or else—which amounts   95
to the same thing—reducing the person to life alone. From this point of
view, no *essential* distinction can be established between humans and
animals. Now, what is in fact missing in animal 'action' is this possibility
of 'deepening', this ability to make breaks and to oppose, in a word this
vertical dimension—which is also the dimension of the intellect—that

appears as both the particular attribute of humans and the indelible quality of the real present: *even 'in' time, humans remain upright*.[3]

This line running perpendicular to the temporal axis of life alone in fact gives consciousness of the present the means to flee and escape, to expand and deepen which have very often led to the present instant being likened to an eternity.[4]

Work by Straus and Gebsattel, to which Minkowski has so effectively drawn attention, clearly shows certain consequences of this temporal superimposition. Basing himself on Höningswald's distinction between immanent time and transitive time, or more simply between the time of the self and the time of the world, Minkowski establishes the duality of sequences as well as the very variable relations of dependence going from one time to the other. Even in normal life:

> there may be discord between them. Sometimes the time of the self seems to go faster than the time of the world and we have the impression that time is passing quickly, life smiles at us and we are full of joy; sometimes on the contrary the time of the self seems to run more slowly than that of the world, time therefore drags on forever, we are gloomy and world-weariness takes hold of us.[5]

Were we to see this as but a trite analysis of the feeling of languor 96 that makes us 'find time too long', we would not have fully grasped Minkowski's intuition. We are not in fact dealing with an *illusion* here but with a psychological reality evident in the analysis of pathological cases. Thus, in some states of endogenous depression:

> the contrast between the two modes of time becomes striking. Here, immanent time seems to go markedly more slowly and even to stop, and this modification of temporal structure comes to be inserted between on the one hand the underlying biological disturbance and on the other the current clinical symptoms; in Straus's view, this modification is the direct consequence of the biological disturbance, which here consists of an inhibition.

It seems that in some way, such patients are disconnecting. They make a perpendicular escape from the duration of the world. In order to set immanent time in motion, particular rhythms of transitive time are then needed. The case of a woman patient of Straus's is very instructive in this respect, for she 'only felt time move forward when she was doing her knitting'.

IV

Let us now offer an example of superimposed time taken from our own experience; it comes from a dream in whose structure we can distinguish between the different kinds of superimposed time and the parts they each play. I had bought a house, and I fell asleep one night thinking of some of the things still to be done. In my dream, my continuing worries meant that I met the owner of my old home. I took the chance therefore of telling him about my new acquisition. I spoke kindly as I was about to give him a piece of bad news; could anyone fail to regret the loss of a philosopher-tenant, one who is ever content and uncomplaining, who combines all the integrity of a moral principle with a hermit's frugality! Then slowly, and with a skill that revealed the striking continuity of capitalist time within me, of which I was entirely unaware, I suggested to my landlord all the many ways in which we might mutually agree to end the contract binding us. I spoke at some length, with sweet words of 97 courtesy and persuasion. My little speech was well organised: the fact that my aim was clear meant that my arguments were produced at exactly the right moment. Suddenly, I looked at the person to whom I was speaking; he was now listening to me very calmly, and he was not my landlord. He had certainly been my landlord to begin with, this I realised through some strange kind of recurrence; he had then been my landlord in his younger days, and afterwards had turned into someone progressively more different until I suddenly realised that I was telling all this to a complete stranger. This piece of bungling on my part annoyed me so much that I flew into a temper over this fresh evidence of my absentmindedness and of the temporal discords that I had allowed to occur within me as a result of my having 'super-imposed time'. I was awoken by the anger that so often, in our dreams, disrupts and shatters time.

Do we need any further proof that verbal time and visual time are in fact only superimposed and that, in all our dreams, they are independent of each other? Visual time passes more swiftly, and it is for this reason that they fall out of step. Had I been able not only to free myself of my financial worries but also to speed up what I was saying, I would have maintained complete synchronism with what was happening visually.

Dreams are indeed extremely changeable horizontally, that is to say along the plane of the normal, everyday incidents of life, yet even so, my dream would at least have retained its vertical coherence, that is to say the form of normal, everyday coincidence. In my conversation with the stranger who took my landlord's place, I would have chosen words which were *appropriate*. I would not have *continued with* my story: I would have *modified* my confidence the moment my confidant began to change.

If we agree to analyse complex dreams from the standpoint of these differences in temporal rate, we shall see that there is much to be gained from the concept of superimposed time. Many dreams will seem incoherent simply because there is a sudden, instantaneous loss of co-ordination between the different times that we are experiencing. It would appear that when we sleep, our different nerve centres pursue 98 their own autonomous development and that they are in effect time detectors, each with its independent rhythm. Let us say in passing that these isolated detectors are particularly sensitive to temporal parasites. Indeed, often in the peaceful repose sleep brings I have the feeling that parts of my brain are crackling, as if the cells were exploding or some kind of partial death were rehearsing its disasters. If we consider time in the context of cellular activity, we must see that it is closer to the time of a moth or an amoeba, any coincidences there being exceptional. When like a beehive, the whole of our brain comes to life; it is statistical time that restores both regularity and slowness. Moreover in waking life, reality offers grounds for agreement. Reality makes what we see wait for what we say, and as a result of this we have objectively coherent thought, a simple super-imposition of two terms which mutually confirm one another and usually suffice to give an impression of objectivity. We then say what we see; we think what we say: time is truly vertical and yet it also flows along its horizontal course, bearing with it the different forms of our psychic duration, all according to the same rhythm. Dreaming is the very reverse of this, for it disengages these different kinds of superimposed time.

V

We have probably now adduced sufficient evidence, evidence drawn moreover from sufficiently heterogeneous sources, to have some kind of certainty that with this temporal superimposition, we are touching upon a natural problem. Let us suggest then how we personally would wish to direct research in order to solve this particular problem.

The temporal axis that lies perpendicular to transitive time, to the time of the world and of matter, is an axis along which the self can develop a formal activity. It can be explored if we free ourselves from the matter that makes up the self and from the self's historical experience, in order to consolidate and sustain aspects of the self which are progressively more formal, and which are indeed the truly philosophical experiences of that self. The most general and the most metaphysical method of approach would be to build up tiers of different kinds of cogito. We shall return later to particular examples of this that are   99 closer to everyday psychology. Let us now turn without further delay to this attempt to create a compound metaphysics, a compound idealism, which will put in the place of *I think, therefore I am* the affirmation that *I think that I think, therefore I am*. We can see even now that *existence* as it is averred by the *cogito cogitem* will be much more formal than existence as it is implied by thought alone; if eventually we can manage to reveal what we really are when we first establish ourselves in the *I think that I think*, we shall be less inclined to say that we are 'a thing which doubts, understands, conceives, affirms, denies, wills, refuses, which also imagines and feels'.[6] We shall thus avoid settling into a phenomenal existence which needs permanence in order to be confirmed. The Cartesian cogito is necessarily discursive for it is entirely horizontal, and this fact has been made abundantly clear by Teissier du Cros in an article that we consider to be quite unusually profound. He argues that:

> between the *I* and the *am*, there is the relation of affirmation and confirmation. Where the self is concerned, the judgement of existence is, in the end, a *repetition*: if we take both of them from the same standpoint, that of realities, and compare the specific experience of the self with the specific experience of things, it will be said to be the same thing as this.[7]

If however we can rise to the *I think that I think*, we shall already be free of phenomenological description. If, continuing a little further, we reach the *I think that I think that I think*, which will be denoted by $(cogito)^3$, then separate, consecutive existences will appear in all their formalising power. We have now embarked upon a noumenological description which, with a little practice, will be shown to be exactly summable in the present instant and which, by virtue of these formal coincidences, offers us the very first adumbration of vertical time.

What we are doing here is not in fact thinking ourselves thinking *something*, but rather thinking ourselves as *someone* who is thinking. Indeed, with this formalising activity, we watch the person being born. This formal personalisation takes place along an axis whose direction is entirely opposite to substantial personality, the personality that is supposedly original and deep, although in reality it is encumbered and weighed down by passion and instinct, and imprisoned by transitive time. Along the vertical axis we are suggesting, being is made mind in accordance with the degree of its awareness of this formal activity, of the power of the cogito it is using, and also of the highest exponent of the compound cogito to which it can go in its attempted liberation. Were we to overcome the difficulties surrounding the first severance, and then reach for example $(cogito)^3$ or $(cogito)^4$, we should immediately recognise the great value of this strictly tautological psychology in which being is really and truly self-concerned, that is to say the value of repose. Here, thought would rest upon itself alone. *I think the I think* would become the *I think the I*, this being synonymous with *I am the I.* This tautology is a guarantee of instantaneity.

It will however be asked in what way this sequence of forms can acquire a specific temporal character. It can do so because it is a becoming. This becoming is doubtless peripheral to the becoming of things and independent of material becoming. Clearly then, this formal becoming rises above and overhangs the present instant; it is latent in every instant that we live; it can shoot up like a rocket, high above the world and nature, high above ordinary psychic life. This latency is an ordered succession. Any alteration in the order of the various tiers is inconceivable. It is, we can be sure, a *dimension* of the mind.

Someone is bound to ask whether this dimension is infinite. To draw such a conclusion would be to yield far too quickly to the seductions of logic and grammar. We therefore refuse to go on lining up our subjunctives indefinitely. In particular, we refuse to imitate those writers who talk so vaguely about *knowledge of knowledge* . . . precisely because the subjective factor of formalisation is not always clearly implicit in *knowledge of knowledge* . . . , in $(knowledge)^n$. We ourselves have found it exceedingly difficult, psychologically speaking, to attain to $(cogito)^4$. We believe that the true region of formal repose in which we would

100

101

gladly remain is that of (cogito)[3]. From the researches into compound psychology which we shall be outlining later on, we shall see that the power of three corresponds to a state that is sufficiently new for us to have to exert ourselves considerably and for a long time there before we can go beyond it and proceed with our construction. (Cogito)[3] is the first really unballasted state in which consciousness of formal life brings us a special kind of happiness.

These different temporal levels can, in our view, be rather schematically and crudely characterised by a number of different mental causalities. Thus, we consider that if (cogito)[1] is implied by efficient causes, then (cogito)[2] can be ascribed to final causes, since if we act with an end in view, then we are acting with a thought in view while being at the same time conscious that we are thinking that thought. Only with (cogito)[3] will formal causality appear in all its purity. This division into things, ends, and forms will of course seem artificial to any linear psychology that seeks to place all entities on the same level, making them part of a single reality, beyond which there can only be dreams and illusions. Yet the discursive, hierarchical idealism for which we are arguing is not limited to this one realist point of view.[8] If we take Schopenhauer's fundamental axiom as our starting point and say that the world is my representation, then it seems acceptable to attribute ends to the *representation of representation* while forms that are constituted in those mental activities which imply both things and ends must be attributed to the *representation of the representation of representation*. Psychologically speaking, if we follow the axis of liberation and free ourselves of all things material, we shall then no longer determine our own being by referring to things or even to thoughts, but rather by reference to the form of a thought. The life of the mind will become pure aesthetics.

Finally, the time of the person, vertical time, is plainly discontinu-   102
ous. Were we to attempt to describe the passage from one power of the cogito to the next as a continuous process, we should realise that we were placing it along the familiar axis of time, and by this we mean vulgar time. This would lead to a complete misunderstanding of temporal superimposition. We would be starting from the mistaken belief that all psychological analysis is necessarily temporal, in other words that all psychological description is historical, and that it is because we are following the hands of the clock that we can successively *think*, then *think that we think*, and then *think that we think that we think*. We would be

disregarding the principle of the fundamental instantaneity of all well-ordered formalisations. If we take psychological coincidences not simply in the instant but also in their hierarchical form, we shall receive far more from them than potential linear development. We remain entirely convinced that the mind thrusts far beyond the line of life.

Let us live temporally at the power of three, at the level of cogito cubed. If this third state is examined temporally in relation to the first state, that is to say in relation to transitive time, it will be full of lacunae, broken up by great intervals of time. Here, quite unmistakably, we have the dialectic of time. And once again, continuity must be sought elsewhere, in life perhaps or perhaps in primary thought. Yet life and primary thought are so devoid of interest for anyone grown familiar with that formal state in which we seek repose from life and thought that as a result, this purely material continuity will pass unnoticed. What we need then is some kind of rational coherence to replace material cohesion. In other words, if we wish a purely aesthetic thought to be constituted, we must transcend the temporal dialectic by means of forms, by means of the attraction of one form to another. Were we to retain our ties with ordinary life and thought, this purely aesthetic activity would 103 be entirely fortuitous, lacking any coherence or any 'duration'. In order that we may have duration at the third power of the cogito we must therefore seek reasons for restoring the forms we have glimpsed. This will come about only if we can teach ourselves to formalise a wide variety of psychological attitudes. We shall now outline some applications of this compound psychology, emphasising the homogeneity of certain textures of time that are very full of lacunae.

VI

Let us first consider an intellectual attitude in which there are a large number of periods of inhibition and few really positive actions. Let us for example study the temporal texture of pretence and see that this has already become detached from the continuous pattern of life: pretence is already a temporal superimposition. Indeed, even when we first consider it, we cannot but be struck by the fact that the texture of pretence is full of lacunae. A continuous pretence cannot really be imagined. Moreover, we must not exaggerate if we wish to pretend successfully. The principle of necessary and sufficient reason is carefully applied when we pretend, with the result that we try to balance inhibitions and actions. Pretence curbs natural expansiveness and cuts it short; it obviously has less density than a feeling that comes straight from the heart. Pretence is no doubt inclined to make up for what it lacks in number by intensity. It reinforces certain characteristics and increases susceptibilities. It brings both invariability and inflexibility to attitudes that are naturally more mobile and more flexible. In short, the temporal texture of pretence is both full of lacunae and also uneven.

If we are to pretend successfully, we must indeed make what is essentially discontinuous and disparate appear to be continuous. The density and regularity of its temporal texture must be increased or, as Dupréel would say, this texture must be consolidated. Acting appropriately does not suffice for this to happen. It only leads us to use circumstance, to constitute from the standpoint of worldly convention and with the world's time an emotional form that cannot really be described as 104 psychological 'consolidated'. Successful and active pretence that is no longer fortuitous requires that it be incorporated into the 'time of the self'. For it to be truly constituted the following paradox must be resolved: pretence has to be attached to the 'time of sincerity', to the time of the person almost to the point when we are deceived by our own deception. It is in this way, in fact, that some pretences at neurosis do in reality come about. More simply, it is by attaching them to the 'time of the person' that we can pretend to feel those false rushes of enthusiasm that sweep others along synchronically with our dynamism. For lying to have its full effect, the different kinds of personal time must as it were mesh with one another. In the absence of this application on to our own rhythm, pretence cannot be given dynamic conviction.

These observations may well seem both superficial and artificial. Where the psychology of an attitude as precise as pretence is concerned, people want psychologists to depict a particular kind of pretence and not 'pretence in itself', and especially they want psychologists to describe to them how truth is turned into falsity and also to let them experience the ambiguity of meaning. We ourselves though are looking for grounds for establishing an abstract psychology and thus in our view, it is precisely because meaning is ambiguous that we can cut ourselves off from it more effectively; pretence seems to us a good example of abstract psychology, of psychology that is formal and artificial, in which time will show itself to be an important characteristic. Indeed, if we take away the double meaning of pretence and consider neither what we are pretending nor why we are pretending, then what is left? Many things: there is still the order, place, density, and regularity of instants in which the person who is pretending decides to exaggerate nature. The schema of initiating acts is all the more important here because it is artificial. The purely temporal aspect of deceit must catch the attention of deceivers themselves. Those who pretend must remember they are pretending. They must feed their pretence. Nothing urges them on or forces them 105 to act, but they still know that it is once again time to pretend. Missing a chance to pretend sometimes—not always—means that pretence is shattered. However full of lacunae it may be, pretence would by virtue of this partial forgetting lose its 'continuity', proving fairly clearly that there can be 'continuity' without a real continuum. In the context of the artificial feeling of pretence, continuity does not need the entirely natural continuity of life that natural feelings have.

When we arrange all that can connect us to others in a close series, when we carefully adapt to fit the time of other people and if at all possible predict others' capriciousness, then all this does not require that we and others share the same substance. Enabling us to share the same time is now though one of interpsychology's main aims. Once this synchronism has been achieved, that is to say when correspondence has been established between two superimpositions of two different psyches, we realise that here we have nearly all the substitutes for a sharing of substance. The time of thought marks thought deeply. We may not perhaps be thinking the same thing but we are thinking something *at the same time*. What an amazing union! All interpsychology ought first to pose the problem of temporal correspondence and not take synchronism to be an effect without discussing it. It is often a convention; it is sometimes calculated; it can always be something that is well made and economically organised. In any case, for artificial feelings, for all feelings we pretend to have, the problem of synchronism is in our view a primary one: time must not be allowed to destroy the work of time. Nor indeed must time be forced.

With pretence, we have seen an attitude which is sustained in a time full of lacunae and already very free of all the obligations of the time of life, a time that is in a way superimposed on the time of life. To give a better idea of our dialectical stance and of the importance of inhibitions that intervene and refuse life's suggestions and connections, let us ask ourselves whether we could by increasing the actions of 106 inhibition reach attitudes which are ever more full of lacunae, in times that are superimposed on one another. Could we for example pretend to pretend, and if the answer is yes, then what would be the temporal form corresponding to pretence at pretence which we shall denote as (pretence)^2?

It would not be difficult to find a good number of literary texts to show that pretence at pretence has not escaped novelists' attention. George Sand refers to it explicitly in chapter 13 of her *Horace*. We would find that it has left its mark on very many pages of Dostoyevsky's work, to the extent that the question arises as to whether his psychology is not a systematically 'compound' one, a psychology that reflects back on itself and is made up of feelings which are raised to 'exponents'. We have only to reread *Crime and Punishment* in particular to see many examples of (pretence)^2, and if we agree to use the schemas of temporal analysis suggested here we realise that these schemas can bring out characteristic features. Thus, (pretence)^2 will appear as much fuller of lacunae than pretence pure and simple. This will be seen if we make even the very slightest attempt at statistics and compare among the instants of pretence those that go from (pretence)^1 to (pretence)^2.

Yet the problem is not of course just a problem of literary psychology. In speaking to different people—to women especially—about the pretence of pretence, we have been surprised at how quickly they understood us. The question as to whether one can pretend to pretend elicited the immediate response: but of course. On the contrary, as soon as we asked the question whether you can pretend to pretend to pretend, everything grew confused and brought on a kind of mental vertigo. Simply because of this confusion, (pretence)^3 poses an interesting problem for compound psychology and temporal superimposition. However hard it may in fact be to establish oneself in this very unstable state, we believe that with a little experience it can be studied. We must not of course put our trust in a purely verbal process and imagine it is enough to give a name to a state for it to be known. For anyone who thinks this can be done, it will be an easy matter to define (pretences)^4, (pretences)^5 and so on. In our own personal experience, we have never really been able to go beyond (pretence)^3. Pretences that go beyond this seem to us to go through grammatical intermediaries that have no psychological value. They cannot in our opinion become temporal in the sense we shall shortly explain.

Having rejected states with too high an exponent, we must now respond to objections we have come across from those who refuse to accept the psychological reality of psychology to the power of three. (Pretence)^3 is often attacked using the objection that since (pretence)^2 is already a return to what is natural, (pretence)^3 is therefore just pretence. Objections like these amount to psychology being associated with logic. Pretence is seen as related to definite truths and two negations are too quickly thought to make one affirmation. Once we have freed ourselves from these automatic inversions and reached real psychological inversions, then many interacting nuances appear and give sufficient pretexts for diversity. We had only just finished giving a lecture on (pretence)^3 when a number of those present were kind enough to give us some interesting notes. One set of notes from Mr. L. Thiblot seems so clear to us that we reproduce it here without making any changes:

A first hypothesis. Simple pretence. A lecturer's class is boring me stiff. But as I am keen for this lecturer to have a good opinion of me, I simulate great attention while he is speaking. I hope he will be duped by my pretence.

A second hypothesis. Pretence to the power of two. The lecturer's class is boring me stiff and as I have reasons for wanting to be nasty to this lecturer, I simulate such exaggerated attention and enthusiasm that he is obliged to say to himself 'this is too good to be true; this student is making fun of me!'. I therefore only pretend to pretend. I pretend but hope the lecturer will not be duped by my pretence.

A third hypothesis. Pretence to the power of three. I am finding the lecturer's class very interesting. Yet as I have had a bet with other students that I shall be nasty to him, I want him to think his class is of no interest to me. I use exactly the same means I have just described in order to do this. I make the pretence of such exaggerated attention and enthusiasm that he is obliged to take them by antiphrasis, as it were. Here there is pretence to the power of three: I *give a semblance* of working so as to *make a pretence* of having a feeling (the lack of interest is itself only a *sham*).

108

If moreover we consider the problem from a temporal point of view, we shall see that the accusation of being merely logical artifice does not hold water. Indeed two negations would make one affirmation if all the first states were to be transposed. This would be the case if we only had one temporal level, one single texture, with everywhere the same continuity. But just as (pretence)^2 is more full of lacunae than (pretence)^1, so (pretence)^3 has even more lacunae than (pretence)^2. In order for the

influence of the rare and chosen instant to be fully understood, let us
adopt a totally analytical approach that should help us learn the art of
pretending to pretend to pretend. Since everyone has knowledge of pretence at pretence, let us entrust this (pretence)^2 to speech and then ask
the eyes to take charge of (pretence)^3. This they do, as they wink or flash
at the right moment. Here again we see the same temporal dissociation,
only this time deliberate, that we pointed out with regard to a dream of
ours. Superimposed times can each be consolidated by particular kinds
of behaviour in which different emotional processes may be involved.

Lastly, those present at this lecture made other suggestions, most   109
of which amounted to bringing more and more speakers into play. We
would thus have the opportunity of varying our social times as much
as we wished, attaching a time to each distinct society. Every state of
pretending would be determined by one particular witness. A would not
be the same for B as for C or D. Temporal superimpositions would be
easily obtained but they would not really be in a hierarchy. Finally, we do
not accept those different pyramidal constructions that are all too easily
made; we ourselves return again to an entirely temporal superimposition
in which emotions that come into some kind of combination with each
other appear as real 'formalisations'. It is a process that can only be fully
clarified by real reflection in which form recognises its independence
from matter. Then the temporal schema truly marks the form and appears
as a characteristic aspect of the psychological element envisaged.

VII

Many other psychological compositions could of course be studied: the
joy of joy, the love of love, the desire for desire, and many more superimpositions too, a large number of examples of which could be found
in the contemporary philosophy of emotion. In particular, it seems to
us that a study of Paul Valéry's work from this standpoint would be
fruitful. Jean de Latour's fine book on him gives rethought values,
re-evaluated values, and re-formed forms their rightful place.[9] This
truly is the dynamic secret of Paul Valéry's active idealism.

In such psychological compositions, difficulties will again appear
starting from the power of three; we in fact attain pure idealism starting
from the power of three. Thus in (love)^3 the ever fickle, the systemati-   110
cally fickle pleasure of (love)^2 is seen to disappear. Moreover, this (love)^2

is still involved in varieties of (love)[1]. Adherence to the *object* only disappears with (love)[3] which is at last free and faithful, and the pure art of love.

Our task though is not to make a thorough study of exponential psychology, and these brief notes are only intended as suggestions for future work. What we would like to draw attention to in concluding this discussion is the value for those conducting work of this kind of taking temporal characteristics as their starting-point. This is where we ourselves would begin such a study: it is obvious that attitudes at the power of two are temporally more full of lacunae than are primary attitudes. Generally speaking, when coefficients are raised we move into times which are increasingly full of lacunae. We believe that despite these increasing voids, a psyche can maintain itself in exponential attitudes without being dependent on the primary psyche. *Idealised times are therefore consistent without however having continuity.* This is one of the principal arguments of the philosophy of time that we are proposing. It would no doubt seem simpler to postulate the continuity of the primary attitude as being fundamental and to consider escapes from this as like rockets which are separate from it, shooting up from time to time along the length of natural development. But this simplest of solutions is not ours. It does not take into account the fact that some minds can maintain themselves in exponential thought, in thought of thought for example and even in (thought)[3]. It therefore seems to us that time in the second or third superimposition has its own grounds for making a sequence. Everything we have already said about psychological causalities being regarded as different from physiological causality could be repeated here in order to prove that reasons and forms can stabilise attitudes which do not really rest on any deep foundation. In superimposed temporal developments we see that, if we examine the higher lines of our mind, events that are very infrequent indeed do suffice to sustain a mental life and to propagate a form. Unfortunately psychologists have no interest in working in this field—'up in the clouds', as malicious critics will say. Contemporary psychology prefers to follow Freud in his acherontic explorations, seeking to feel thought at the sources of life and from the standpoint of life's urgent flow. Even though pure thought is revealed as clearly discontinuous while at the same time remarkably homogeneous, psychologists still want the psyche to be a form which is equivalent to life and always contemporaneous with a development of life. And yet the more full of lacunae a psyche is, the clearer it is;

111

the more concise its orders, the more they are obeyed. Times that are truly active are times that have been emptied out in which conditions of execution only appear as subordinate conditions. When we have studied artificial psychology and exponential attitudes, we shall see that the times of action are isolated and that their repetition is not totally conditioned by execution but indeed first of all by higher and more mental necessities. The coherence of reasons for acting will determine the cohesion of real actions. Continuity at the higher levels of time will become a metaphor. It will be all the clearer for this, more thought-provoking too and finally more easily restored.

This admission that continuity is a metaphor should not, in our view, be seen as an objection to our argument since this is in fact the case for all durations. To prove this, we shall now study some of the most common metaphors serving to depict the constant action of duration. We shall see from these metaphors that continuity is always bound up with a point of view, or in other words that it is purely and simply a metaphor.

NOTES

1. Bachelard's footnote: Lecomte du Nouy, *Le Temps et la vie* (Paris: Gallimard, 1936); see in particular chapter 9.

2. Bachelard's footnote (amended): A. Koyré, 'Hegel à Iéna', *Revue d'histoire et de philosophie religieuses* (1935).

3. Bachelard's footnote (amended): A. Marc, '*Le Temps et la personne*', *Recherches philosophiques*, 4 (1934–1935), 132; for information regarding this journal, see chapter 5, note 3.

4. Bachelard's footnote (amended): cf. A. Rivaud, 'Remarques sur la durée', *Recherches philosophiques*, 3 (1933–1934), 19ff.

5. Bachelard's footnote: E. Minkowski, *Le Temps vécu* (Paris, 1933), p. 278.

6. Haldane and Ross, *The Philosophical Works of Descartes* (New York, 1955), 1, p. 153. Bachelard refers here to Descartes's *Méditation II*.

7. Bachelard's footnote: Ch. Teissier du Cros, 'La Répétition, rythme de l'âme, et la foi chrétienne', *Etudes théologiques et religieuses* (Montpellier, May 1935).

8. There is an implicit reference here to Bachelard's article on discursive idealism, 'Idéalisme discursif', first published in *Recherches philosophiques*, 4 (1934–1935), 21–29, and subsequently published in G. Bachelard, *Etudes* (Paris, 1970), pp. 87–97.

9. Bachelard's footnote (amended): J. de Latour, *Examen de Paul Valéry* (Paris: Gallimard, 1935).

Chapter 7

Metaphors of Duration

I

If readers have followed our argument that connections between truly   112
active instants are always made at a level different from that at which
action is executed, they will not be far from concluding as we do that,
strictly speaking, duration is a *metaphor*. This makes the ease of illustration that is one of the charms of Bergson's philosophy much less
surprising. Indeed, it is not in the least surprising that metaphors can be
found that illustrate time if we make them the single connecting factor
in the most varied of domains, in life, music, thought, emotion, and history. We think that by superimposing all these more or less empty, more
or less blank images, we can make contact with the fullness of time and
the *reality* of time; from a blank, abstract duration in which just the
possibilities of being would be found, lined up one after the other, we
think that we can move on to duration that is lived, felt, loved, sung, and
written about in literature. Let us again outline these superimpositions:
when duration is regarded as life, it is the inseparability and organisation of a succession of functions—life when continuously conscious of
something is reverie—reverie itself is a melody of the mind, its con-
stituents paradoxically both free and merged together. If we go on to
add that in the same way melody 'can be compared to a living being',[1]
we have created a whole family, an entire closed cycle of metaphors   113
that will constitute the language of continuity, the song and indeed the
lullaby of continuity. Tranquil duration, life that is well balanced, music

109

that sweeps us along, sweet reverie, clear and fruitful thought, all of these offer us experiences 'proving' time to be continuous. These are all happy experiences: duration is a synonym of happiness or at the very least, the synonym of a possession, a gift. The clear fact of possession helps to uphold the promise of duration.

There is just one snag in all this: no experience is self-sufficient; no temporal experience is really pure. We need only look closely at any one of the images of continuity and we shall always see the hatched lines of discontinuity present there. These hatched lines cast a continuous shadow only because of heterogeneities that have become blurred. This is an argument that we have already put forward several times. We shall give it a new direction here by examining one particular metaphor and attempting to analyse the *density* of music and poetry. With music for example, we shall have to show that what makes for continuity is always a hidden dialectic evoking emotions with regard to impressions, and memories with regard to sensations. In other words, we shall need to prove that the continuum of melody and also that of poetry are emotional reconstructions which come together beyond the real sensation, thanks to emotion's vagueness and torpor and to the hotchpotch of memories and hopes; as a result of this, they come together on very different levels from the one we would be confined to by a scientific study of contextures made of sound alone.[2]

Let us first emphasise this reflux of impression flowing from the present to the past and giving to rhythm, melody, and poetry the continuity and life they lacked when first produced. A lapse of our attention to melody would be enough to end this reflux. Successive notes then no longer *sing* but remain in the qualitative and quantitative discontinuity in which they are produced. Sensations are not connected; it is our soul that connects them.

114

The continuity of the fabric of sound is so fragile that *a break in one place sometimes causes a break in another*. In other words, step by step connection does not suffice; this partial connection is dependent on a loose-knit association, on the continuity of the whole. We must in fact *learn* the continuity of a melody. It is not *heard* straightaway, and it is often the recognition of a theme that makes us aware of melodic continuity. Here as elsewhere, recognition takes place before cognition. Lionel Landry has rightly said that 'a rhythmic pattern does not take on its full qualitative value for the person hearing it just once'.[3] Initially when sounds first developed, temporal structure was not really *formed*; the causality of music was not yet established. Structure and causality were postulated in the domain of possibility rather than in that of reality. And everything remained disconnected and gratuitous. Then comes the recurrence of impression, bringing formal causality. For a metaphysician, this formal causality corresponds to that *qualitative value* referred to by Landry.

This reform that does indeed give form can create symmetries in poetry and music from a starting-point in subordinate dissymmetrical forms. This is what Raoul de la Grasserie has noted: 'two lines follow each other and I assume that within each of them, there is in the two halves of the line an unequal number of syllables; if this inequality is reproduced in the second line and in the same sense, then the same rhythmic pattern will reform, *internal inequality* having now become *external equality*'.[4] In other words, the identity of the complex entity will transcend the diversity of detail; something will in a way be completed by their symmetry. It is thanks to the group that there will be continuity. Thus poetry, or to be more general melody, *has duration* because it *begins again*. Melody duets dialectically with itself, losing itself so that it can find itself again, knowing it will be absorbed in its first theme.[5] In this way then, it gives us not really duration but the illusion of duration. In some respects, melody is a kind of temporal perfidy. While it promised us development, it keeps us firmly within a state. It takes us back to its beginning and in doing so, gives us the impression that we ought to have predicted where it was going. Yet strictly speaking, melody does not have a primary source, a central point from which it spreads out. Its origin is revealed by recurrence and just like its continuity, this origin is a composition.

If we now examine this dialectical deletion of the first theme, we shall be persuaded that a *repeat* cannot really be seen as melodically linked to the effect it first had on us. Between one refrain and the next there is less than a latent memory, and less still than a very particular expectation. Expectation is never as plainly negative as it is in music, for it will in fact only become conscious expectation if the phrase we have heard is repeated. We shall not remember having expected it; we shall simply recognise that we ought to have expected it. Thus, what gives melody its light, free continuity is this wholly virtual expectation which is real only in retrospect, and just a risk to be run, a possibility. As Maurice Ravel once said, 'comparisons with architecture are absolutely pointless: there are rules for making a building hold together but

115

116

none for linking musical transitions'.[6] In reality, this linking is upheld by extramusical intermediaries, by emotional, dramatic, even literary values. Were we to stem the flood of emotion accompanying melody, we would realise that melody as just a sense datum would cease to flow. Continuity does not belong to the melodic line itself. What gives this line consistency is an emotion more vague and viscous than sensation is. Music's action is discontinuous; it is our emotional resonance that gives it continuity.

Music's emotions are thus an attempt at a temporal synthesis, an attempt which is never fully realised because music's causality is always deferred, always systematically deferred. Its action is not one that advances by degrees. Raoul de la Grasserie has clearly understood the importance of this causal deferment which is fundamental to what he calls *discordant harmony*:

> Harmony in music is not always achieved immediately; in modern music especially, harmony is often delayed for a while so that it produces a much greater effect after it has been waited for. A note is produced, then another follows it; if we stopped there, there would be absolute discord, music that is out of tune, and an absence of rhythm; the ear is not yet injured but it is already anxious, it is in pain and feels something similar to what at a lower level is the sensation of hunger. Were this state to go on too long, exasperation would set in; the musician acts in time though, producing the note that *resolves the discord in a final chord*, in a harmony both desired and sought, and because of this all the more astonishing.[7]

117

We thus place drama above the sound, and the unity of this drama—a unity we understand in retrospect—makes the melody flow backwards and brings continuity to sensations first felt in their more or less complete isolation. We then go over the whole page again, restoring the finality of music that really does provide the only possible proof of melodic causality; in this way we attain 'that special, purely musical tranquility which transcends both heaviness of mind and also sleep; this repose that music produces comes when disymmetries open elsewhere are closed, made now into symmetries'.[8] To sum up then, the impression of fullness and continuity given by music is due to the mixture of feelings it evokes in us. Once we have observed melody's exact relation to time, we realise that embroidery distorts the canvas and that as a result of this, music is a metaphor which is often misleading for a metaphysical study of duration. We shall be persuaded that this is so if we now turn to consider the very perceptive studies written by Maurice Emmanuel.

II

He is a technical expert who, in his book on the history of musical language, has no hesitation in refusing to give primacy to techniques of measurement, that is say to techniques relying simply and solely on objective ways of measuring time. In his view, only what is written down can be regarded as measurable, proving that precise duration is not the essential *substance of music*. Measurement was initially a representation more to do with mnemonics than reality. In modern techniques, measurement allows us to 'read and directly express rhythm's movement and pace'.[9] Metronomes are crude instruments, however. They are the magnifying glasses with which weavers count the threads, not the looms themselves. Metronomes do not even describe the 118 fabric of time very successfully. They cannot control the music that flows from inspiration, new, fresh, ethereal, made only of movement. Emmanuel shows the exaggerated role attributed to the bar line: we must, he says 'close the door in its face when it seeks to enter the sanctuary of rhythm. Its job is just a lowly one; it works like a metronome, regularly marking out the way, and it has no right to claim it represents the landscape any more than a milestone does' (*Histoire de la langue musicale*, 2, p. 442). He gives examples in which fine anapaests[10] are 'hacked to bits' by bar lines. In the present period itself, 'the bar line, having become an indispensable aid to polyphony, does not in any way indicate the rhythm, nor is it in any way linked to it; the constituent parts of the rhythm only rarely correspond to the dividing spaces of the bar lines' (p. 563).

In Lionel Landry's book which is so subtle and so very far from fixed and preconceived ideas, the primacy and intransigence of the absolute temporal framework is also rejected: 'the idea must also be dismissed that there is a primary and indivisible time fundamental to every rhythm. While it is true that rules are found in ancient prosody, we can be sure that beyond the recognised exceptions there are to these rules, variations in pace of delivery were enough to remove their absolute value' (*La Sensibilité musicale*, p. 25). In other words, the temporal relation that gives rhythm a pattern can accept very many deformations.

Moreover, were music a matter of counting up different durations, making it therefore a rigorous chronometry, we would find a new melody by moving backwards through this collection of temporal fragments that have been very carefully shared out. This idea could only occur to someone who transcribes music. As Landry says, 'this proves . . . that the spatialisation of the musical phrase is not natural, it is irreversibility that the temporal flow of music seems to present to us: for example, in fugue, while listeners may easily accept the *inversion* of a theme, they at the same time find *retrogression*, the retrograde movement heard in *canon cancrizans*,[11] artificial, academic, and only perceptible on paper' (p. 29).

Yet once rid of this regular, objective framework of measurement, rhythmic movement will appear in a continuity that is more metaphorical than real. Between rhythm's different movements, the dialectic will be more free and the time of music will, as it develops, be affected by an essential relativity. Thus, all the slow passages are *ad libitum*. They are more subjective than objective. Now, these slow passages are important areas, areas in which deferred emotion is realised. They are periods of melodic relaxation, and in the end there are far more of them than the written page shows. A soul with some expertise in music feels and lives this dialectic of regularity and freedom, of emotion which, deferred and then realised, moves in a wave all along the melody.

If we now go into more detail here, we see that in music, the 'duration' of a note is not the pure and obviously primary element that teachers of music theory would have us believe. Maurice Emmanuel in fact notes that 'in principle . . . intensity is connected to length, in the sense that when there are two unequal elements of duration, it is the longest that is considered *loud*. Length and loudness are related: in primary rhythmics, this is a kind of necessity. In rhythmic versification, loudness calls for length' (*Histoire de la langue musicale*, 1, p. 526). Later, he writes that 'the principle set down by the Ancients was still true in the fifteenth century and will remain so: it is that unless there are particular indications or rules, the relationship established between the duration and intensity of sounds is a direct one' (2, p. 577). For our argument, the fact that this relationship is direct deserves the very greatest attention since it clearly shows intensity as giving duration, and duration as—once again—just a consequence. The melting, fading, indistinct character of melodic connections can thus be derived from the

119

120

momentum imparted by sounds. It is a kind of acoustic penumbra that does not enter into the precise arithmetic of rhythm.

This interference of intensity and duration in melodic phenomena offers an illustration of a theory put forward by Jean Nogué.[12] His theory is based on an ingenious and penetrating study of the energetics of sensation. It amounts to making a distinction in the development of a sensation between correct breath control and the voice's outpouring; it thus allows us to analyse a sensation's both static and dynamic conditions. Were we to compare this analysis with Maurice Emmanuel's discoveries, we would become aware of the way in which the voice takes flight from a starting point in the instant of correct breathing. For it to have duration, the voice needs a reserve of energy. This reserve exists statically before it is expended dynamically. It must be grasped in its initial value for its intensity to be truly measured; the duration resulting from this gives a less accurate measurement of it. The existence of this combination of intensity and duration proves at the very least that duration is not a truly primary quality of music's constituent parts.

This complexity will be clearer still if we become aware that with the dialectic of the long and the short there is intertwined not just the dialectic of the loud and the soft but also that of the high and the low. The atomisation of melody can then be really understood. Lionel Dauriac has carefully marked out the stages in this atomisation. He starts with the 'dyad of the high and the low'. He first of all accepts there is a *continuous* variation from the low to the high. The two pitches will then be connected by 'an inclined plane'. Very soon though, a child's voice playing up and down this 'inclined plane' will turn it into a ladder. Indeed, as he says:

121

> the day a sound that is true is produced in a child's throat, it can be said that the outcome of this random playing of the vocal organ has turned out to be real voice-training. What constitutes this training? It consists of the production of atoms of sound which the new born baby's developing attention separates out in the indefinite field of the low and high. My reason for using the term 'atoms' will soon be understood if it is borne in mind that a sound that is true always remains on the same rung of the musical ladder as long as it lasts, and if it is also remembered that qualitatively, musical sounds are resistant to any variation in degree: however loud or soft the intensity we think they have, a D or a E will always remain a D or an E as long as they reverberate.[13]

At first sight, Dauriac's argument seems to support those who favour a pre-existing continuity, and it will be objected that atomising pitches and timbres is secondary and artificial. However, if we consider this argument carefully, we must note that the 'continuity' which is postulated as being immediate is so ephemeral that it cannot be made a framework for the construction of musical concepts. And vice versa, atomisation takes place so early, it is so spontaneous and so unlearned that in many respects it can be regarded as natural. As Lionel Dauriac himself says, continuity is little more than the 'the seat of confused and incoherent sonorities'.

Thus, if we take a melodic line that is as simple and even as possible, we see the accumulation of principles of atomisation. It would be futile to resist these principles of the phenomenism of sound and to go on seeing the substance of melody in duration. In fact, melody does not 122 provide good metaphors for the psychology of time any more than life does. It would instead give us a wrong idea of time, lending as it does too many intrusive colours to rhythms constructed on the dialectic of sound and silence. It will be better understood once we have made some observations about rhythmic superimpositions.

III

Before discussing the essential relativism of rhythmic superimpositions, we must again banish all habits of reference to absolute time. Here once again, we shall argue the essentially secondary and pragmatic character of measurement. Synchronism is not achieved by an accurate measurement of the different durations but simply by the instantaneous signal given when someone *beats time*. Beating time is, in Expert's view, 'a practical means of carrying out the most arduous superimpositions of disparate rhythms'.[14] It is specious to object that this itself obeys a simple rhythm and that it claims to be bringing in an objective rule, valid for all voices, that is to say a mathematical time with regular durations. Indeed, beating time does not work like duration but like a signal. It brings coincidences close together; it brings together the different rhythms in instants that are always notable ones. Moreover, conductors' actions are far more effective than would be the action of a well-regulated mechanism. They are truly masters of movement and pace rather than dispensers of pure duration. They manage not just duration but breathing, and it is here that the values of intensity take precedence over those of duration. Conductors must often let the sound die away of itself rather than stifle it. They measure the outpouring of the sound in terms of the physical control that first produces it. They also overlay one register on another and control rhythmic correlation.

Here we have an illustration of the paradox mentioned in our fore- 123 word to this book. From the moment we refuse to allow ourselves any reference to an absolute duration, we have to accept fully that *rhythms are overlaid and interdependent*. It would not indeed work if we took there to be one fundamental rhythm to which all the instruments refer. The different instruments in fact support each other and carry each other along. The conductor's role is to make the instrumentalists more conscious of their endeavours to correlate with each other.

Our impression of continuity and fullness stems from this correlation. We do not really know whether it is the quick or the slow rhythm that imparts momentum, precisely because it is co-operation that determines this momentum. We cannot therefore really separate melody from harmony, as Georges Urbain has shown in an article of great density and richness: 'melodic sequence', he writes, 'is completely dependent on harmonic sequence'.[15] There is always something that *accompanies*, that *upholds*. Yet this accompanying and upholding is as insubstantial as that which accompanied and upheld, and it is for this reason that Urbain's paradox can be accepted: 'even when melody is unadorned, that is to say when it is monody', he says, 'there has to be something underlying it and carrying it along; 'harmony is then held to be implied'. It can be said that when we listen to a melody that is as linear as it is possible to be, we give it density, we *accompany* it. We cannot listen to it as a whole without giving it an accompaniment. We cannot hear its connectedness, its continuous duration, without this heterogeneous summation of sound and soul.

Thus, we have once again come to the same conclusion: a homogenous process cannot ever change. Only plurality can have duration, can change, and can become. The becoming of a plurality is as polymorphous as, despite all simplifications, that of a melody is polyphonous. 124 The duration of sound is dialectical in every direction, on the axis of melody as on that of harmony, in intensity as in timbre. Musical metaphors would therefore be far more suited to teaching us the different dialectics of time than to giving us images of the continuity of substance. For this to happen, all we have to do is not rush on to totalisations made by our over-all impressions but choose instead to live, without emotional viscosity, the musical life that really is both full of movement and variation and also free.

IV

The same conclusions could be reached were we approach the study of poetic rhythms in the same analytical spirit. We shall content ourselves with just a few observations in order to show that the rhythmics of poetry gradually breaks away from ideas of measurement and is arithmetised by grouping together notable instants rather by measuring uniform durations.

Ideas of measurement do not appear to have come on the scene at the very beginning. Raoul de la Grasserie has shown that the rhythm of sound alone has come late in poetry. He sees the starting point of prosody as the line:

> which was entirely psychic and formed by the divisions of time between which words, that is to say ideas, were distributed. At this point in development . . . we have the prose of the Bible . . . (Later on), we pass imperceptibly from the same number of words in each phrase, where words are of different lengths, to the same number of syllables, and so the *early form of the line*, the line produced by *counting syllables*, was born (*De l'Elément psychique dans le rythme*, p. 24).

The important point for our argument is the primary character of the *psychic* line in poetry and its inherent supremacy over objective temporal value. We shall rediscover this psychic poetry, this soundless poetry, if we are prepared to *think* the lines instead of scanning them, above even our inner language, in the time of thought, a time that is full of lacunae. We shall then become aware that continuity is essentially dialectical, that it is the result of reconciling opposites and that temporally, it is made up of shifts, movements towards the future or back to the past.

Surrealist poetry would give good examples of this temporal dialectic, this purely psychic rhythm. If it meets with the objections or incomprehension of psychologists given to logical reasoning and also of literary critics, this is because they seek to evaluate this kind of poetry by imposing schemas of continuity on it and do not accept the dialectical freedom on which it is constructed. Beyond sonority and at the level of the new-born psyche, silences may be brief or long. It does not matter which of these they are: we can rest or react, we can allow an impression to fade or else suddenly interrupt it with a different or contrary impression. Then *poetic causality* will appear, in all its disconnectedness; it reverberates for a long while from one centre to another, despite all intermediaries; the waves of syllables are simply eddies. Being a poet means multiplying the temporal dialectic and refusing the easy continuity of sensation and deduction; it means refusing catagenic repose and welcoming a repose that vibrates, a psyche that vibrates.

This thought poetry doubtless needs poetry that is spoken, in which the echo will reveal the voice that lies deep; however, the rhythm we think is our starting-point for organising the rhythm we hear, and not the reverse. As for counting syllables, which is a kind of imprinted rhythm, well, this cannot really be defended. To support our argument here, we need only refer to Pius Servien's very interesting work in recent years on the phenomena of poetic rhythm, work which is in some respects related to Maurice Emmanuel's discoveries. Pius Servien has in fact shown that measurement of duration was very far from forming the basis of poetic rhythm. This measurement of duration would do no more than support an artificial rhythm. As he says in his book *Les Rythmes comme introduction physique à l'esthétique*:

> People have endeavoured to determine long and short syllables very accurately by carefully analysing words, but they did not realise that everything collapses like a house of cards as soon as speech breathes on these fragile edifices. A word's long and short syllables *immediately change form, according to the position of the word in the sentence and the stress on it* (p. 64).

True poetic rhythm is made by grouping tonalities together. It is strengthening; it is intensity; duration is but a more or less faithful consequence. To quote Servien, 'there is only one truly independent rhythmics which rules all the others . . . As secondary rhythmics, that is to say those rhythmics which are totally under the sway of tonic rhythmics, we have timbre first of all and then the different durations'.

A discontinuous Bergsonism could readily accept this *realisation* of tonic groups; however, rhythmic values would of course have to retain the discontinuity of impulses with different intensities, since these discontinuities come together at a very homogeneous level, that at which the phenomenon is observed, all ideas of a deep and hidden life that offers us its fundamental continuum having been abandoned. 'What it is important to measure', according to Servien, 'is the vibration we really hear, and above all the *vibration that is observed*' (p. 27). Now, this does mean eliminating differences that are ineffective, and it also means that the formal cause has supremacy over the material cause. The sound that is produced is nothing in comparison with the sound that is observed. Rhythm will thus be constituted at a level of abstraction at which the mind will very soon play an active role. Servien then makes this very general definition:

> Something can be a rhythmic factor if we can see in it groups of elements   127
> with the following properties. Firstly, the elements of all the groups are
> perceived as being of the same nature; if one of them draws our attention,
> then our attention is led to be interested in all of them. Secondly, the
> elements of the same group appear as equal, and those of two different
> groups as unequal (p. 29).

At this level of abstraction, the precise position of events in a uniform time loses much of its importance, and we realise that the principle of frequency takes precedence over that of measurement. In other words, the question 'how often?' comes before the question 'how long?'. Were people to accuse us here of going in a vicious circle, the objection being raised that in order to compare frequency we have to have equal intervals, our answer would be that tolerance with regard to 'the equality' of intervals is so great that it undermines the whole idea of measurement. Lyric poetry is analysed by looking at the proportion of stressed and un-stressed syllables. When we count up like this we forget duration.

We can understand how Pius Servien came to propose that this kind of generalised rhythmics should be made the basis of all aesthetics. We propose to make it the basis of all temporal metaphysics.

Let us then decide on the fundamental temporal principle of generalised rhythmics: it is the restoration of form. A characteristic is rhythmic if it is restored. It then has duration through an essential dialectic.

If a rhythm clearly determines a characteristic, it will often affect related ones. In restoring a form, a rhythm often restores matter and energy. For example, to quote Servien, 'as it ends, music brings repose to the energies it has created. Usually too, it carries into this repose most of the energies originating elsewhere that it has captured and brought with it' (p. 45). A philosophy of repose can never spend too long reflect- 128 ing on this both formal and occasionalist causality which gives us a very accurate idea of how insistently time presses us. Rhythm really is the only way of disciplining and preserving the most diverse of energies. Rhythm is the basis of the dynamics of both life and the psyche. It is rhythm, not melody—for melody is too complex—that can provide the real metaphors of a dialectical philosophy of duration.

NOTES

1. Bachelard's footnote (amended): H. Bergson, *Essai sur les données immédiates de la conscience* (Paris, 1889), p. 75. For Bergson's use of music as a metaphor of duration here, see his discussion of 'concrete duration' in chapter 2.

2. Bachelard's footnote (amended): Otto has noted the syncretism of Bergson's method, stating that 'Bergson's fluid ideas are in reality the ideograms of feelings and of religious and aesthetic intuitions. In taking them to be scientific ideas, he is confusing ideas with experience, a confusion for which Schiller took Goethe to task' (*Le Sacré*, p.153). Bachelard in fact quotes from the French translation of Rudolf Otto's book, its full title being *Le Sacré, l'élément non-rationnel dans l'idée du divin et sa relation avec le rationnel*; the original text was published in Munich in 1932.

3. Bachelard's footnote (amended): L. Landry, *La Sensibilité musicale, ses éléments, sa formation* (Paris, 1927), p. 29.

4. Bachelard's footnote: R. de la Grasserie, *De l'Elément psychique dans le rythme* (Paris, 1892). It should be remembered here that traditionally, French versification requires syllables to be counted, the Alexandrine for example— used in classical theatre and much subsequent verse—being a line of twelve syllables.

5. Bachelard's footnote: cf. G. Urbain, 'La Mélodie', *Journal de psychologie* (1926), 201. Georges Urbain takes it to be a principle that 'a melodic movement always returns to its origin'.

6. Bachelard's footnote (amended): M. Ravel, *Courrier musical* (1 January 1910); Bachelard states that he has taken this reference from Landry, *La Sensibilité musicale*, p. 185.

7. Bachelard gives no page reference here.

8. Bachelard's footnote: P. Servien, *Les Rythmes comme introduction physique à l'esthétique* (Paris, 1930), p. 45.

9. Bachelard's footnote (amended): M. Emmanuel, *Histoire de la langue musicale* (published in two volumes, Paris, 1911), 1, p. 253. Maurice Emmanuel (1862–1938) was, like Bachelard, a native of Bar-sur-Aube; a distinguished musician—composer, performer, scholar, and writer—he was appointed to a Chair in the history of music created for him at the Collège de France in 1898 but which was soon suppressed due to a faction opposed to this discipline, Emmanuel then being obliged to earn his living teaching music in a girls' *lycée*.

10. In prosody, an anapaest is a foot consisting of two short syllables and one long one (» » -).

11. In *canon cancrizans*, which is sometimes referred to as both retrograde canon and 'crab canon', the melody is given out backwards; the etymology here is rather misleading, *cancrizans* being derived from the Latin *cancer*, the crab with its sideways movement.

12. Bachelard's footnote: a very succinct account of Jean Nogué's theory is found in his excellent article 'Ordre et durée', *Revue philosophique* (July 1932).

13. Bachelard's footnote: L. Dauriac, 'Sur l'origine commune du langage verbal et du langage musical', *Journal de psychologie* (1932), p. 834.

14. Bachelard's footnote (amended): this reference is taken from M. Emmanuel, *Histoire de la langue musicale*, 2, p. 378.

15. Bachelard's footnote: *Journal de psychologie* (1926), 206. He does not give the article's title or page references for his other quotations from it.

matter and radiation are similar. This amounts to saying that like radiation, matter must have wave and rhythmic characteristics. Matter is not spread out in space and indifferent to time; it does not remain totally constant and totally inert in a uniform duration. Nor indeed does it live there like something that wears away and is dispersed. It is not just sensitive to rhythms but it exists, in the fullest sense of the term, on the level of rhythm. The time in which matter develops some of its fragile manifestations is a time that undulates like a wave that has but one uniform way of being: the regularity of its frequency. As soon as the different substantial powers of matter are studied in their detail, these powers present themselves as frequencies. In particular, as soon as we get down to the detail of exchanges of energy between different kinds of chemical matter, these exchanges are seen to take place in a rhythmic way, through the indispensable intermediary of radiations with specific frequencies. Energy that is looked at very generally may no doubt appear to lose its rhythms, letting go of what it has of undulating, wave time; it is thus seen as an overall result, as an overview in which time itself has lost its wave structure. We pay for electricity by the kilowatt-hour and for coal by the hundred-weight, yet we are nonetheless both lit and heated by vibrations. Forms of energy which are still more constant must not delude us. The kinetic theory of gases has taught us that a gas enclosed within a pump maintains the piston at an invariable level through a large number of irregular collisions. It would doubtless not be contradictory that a temporal accord between these collisions might come about, with the piston jumping simply as the effect of these synchronised collisions and without there being any macroscopic reason for it. Physicists have confidence though: the law of large numbers *preserves* its phenomena; the chances of there being a temporal accord between collisions have negligible probability. In a similar way, a kinetic theory of solids would show us that the most stable patterns owe their stability to rhythmic discord. They are the statistical patterns of a temporal disorder, and nothing more than this. Our houses are built with an anarchy of vibrations. We walk on an anarchy of vibrations. We sit down on an anarchy of vibrations. The pyramids of Egypt, whose function is to contemplate the unchanging centuries, are endless cacophonies. A magician who as the conductor of the orchestra of matter could bring material rhythms together, would make all these stones vanish into thin air. The fundamental nature of rhythm for matter is clearly shown by this possibility of there being a purely temporal explosion, due solely to an action that synchronises the superimposed times of the different elements.

Were we to tackle the problem with regard to a particular particle, our conclusion would be the same. If a particle ceased to vibrate, it would cease to be. It is now impossible to conceive the existence of an element of matter without adding to that element a specific frequency. We can therefore say that vibratory energy is the energy of existence. Why then should we not have the right to place vibration at the heart of time in its original form? We do so without any hesitation. For us, this first form of time is time that vibrates. Matter exists in and only in a time that vibrates, and it is because it rests on this time that it has energy even in repose. We would therefore be forgetting a fundamental characteristic if we were to take time to be a principle of *uniformity*. We must ascribe fundamental duality to time since the duality inherent in vibration is its operative attribute. We now understand why Pinheiro dos Santos has no hesitation in writing that 'matter and radiation exist only in and through rhythm' (volume 2, section 1, p. 18). This is not, as is so often the case, a declaration inspired by a mystique of rhythm; it really is a new intuition, firmly based on the principles of modern wave physics.

This being so, the initial problem is not so much to ask how matter vibrates as to ask how vibration can take on material aspects. The theory of the relations between substance and time can therefore be seen in a completely new metaphysical light: it should not be said that substance develops and reveals itself in the form of rhythm, but rather that it is *regular* rhythm which appears in the form of a *specific* material attribute. The material aspect—with the pseudo-riches of its irrationality—is but a confused aspect. Strictly speaking, the material aspect is *realised confusion*. Since chemistry studies not *matter* but *pure substance*, it will sooner or later lead to the definition of the precise qualities of this pure substance as temporal qualities, that is to say as qualities that are wholly characterised by rhythms. In this connection, photochemistry already suggests really new substances which bear the mark of vibrating time. We can predict that chemists will soon produce substances with space-time that has been made symmetrical and rhythmic. In other words, the metaphysician wishing to establish intuitions that are in agreement with current scientific needs must put symmetry-rhythmy in the place of the doubly uniform space-time used in the era preceding de Broglie.[2]

132

It is clear to us that realism does need a metaphysical inversion for it to correspond to the principles of wave materialism. This is a point to which we intend to return in another book in which we shall be able to take all the scientific proof into account. Nor shall we discuss here 133 whether realism thus *inverted* is still strictly speaking realism. All we are doing for the moment is outlining how rhythmanalysis has a basis in physics and showing that this theory, which is really more biological and psychological in fact, stems from a general metaphysical view.

II

We shall be equally brief with regard to Pinheiro dos Santos's proposal of a wave biology. Referring to a large number of facts, drawn for the most part from homeopathy, the author suggests a 'wave' interpretation, that is to say the explanation of the action of substance by substituting for substance a particular kind of radiation. Dilution, which is always very great in homeopathy, in fact favours the vibrating temporalisation of medical substances. This interpretation is plausible, although it does not completely reject the traditional substantialist interpretation. It would doubtless be necessary to set up experiments in discrimination—for example, real medicinal interferences which are conceived in a vibratory way—in order to legitimise fully the wave form Pinheiro dos Santos suggests. Let us simply try to give a metaphysical description of the two opposite and complementary points of view of substance and rhythm.

The usual substantialist intuition is first of all contradicted in a way by the existence of homeopathy. Indeed, for substantialist intuition in its naive form, that is to say in its pure form, a substance acts in proportion to its mass, up to a certain limit at least. It accepts that there may be some very small doses, too much of which would produce disturbances. 134 It does not find it easy to accept the effectiveness of the extreme dilutions used by homeopaths. As long as medical substances are considered to be quantitative realities, it will not be easy to understand the action of substance taking place in, so to speak, *inverse* ratio to its quantity. In the same way, in a rational approach to health-care, food substances are always held to be dependent on an assessment of their weight. The human body is like a food store in which no shelf must remain empty. We must take our daily dose of the different food-stuffs that, matter for matter, must come together again in the body's economy. Here again, quantitative intuition is placed firmly in the foreground.

A psychoanalysis of the feeling of having could be undertaken at this point. Jokes about homeopaths go down very well and this is doubtless because of the predominant pleasure of possession, of possession that is very clearly physical and material, that results from consciousness of digesting and growing. It is against this major and immediate sense of security given by the joy of swallowing that homeopathy and a wave conception of healthcare must react. These theories concerning small dosage do not just have the idea of substance against them but also the obvious feeling of strength we get from possessing a substance, from cherishing our reserves and our capital.

However, against this first reluctance to be convinced, let us accept homeopathy and see how Pinheiro dos Santos interprets it rhythmanalytically. He views assimilation as not so much an exchange of substances but an exchange of energy; since energy in the detail of its development cannot escape vibratory form, Pinheiro dos Santos proposes the systematic introduction of radiation between the substance that is swallowed and the substance that is assimilated. The phrase *assimilated substance* has little meaning, in fact. If this is simply a matter of storing something away, as in the case of adipose cells, then we are not dealing with the anagenetic action of life. It is at the very moment that substance is expended and destroyed that its action must be understood. (We are not saying this should be at the moment that substance is transformed, for wave materialism can postulate the destruction of matter.) Now according to wave biology, it is not possible for substance to act truly if it is not *temporalised* in a vibratory form, following its destruction. When it is stored away, it is shut into inert space. It only acts where it is, that is to say on itself. In order to go beyond itself, it has to be propagated and it can only be propagated in waves. External action is necessarily action that vibrates. Moreover, the intervention of a wave will always be necessary in order to awaken and activate a substance that has been stored away. We must therefore always return to the period of activation in order to understand the action of a food-stuff or a medicine.

Hence, therapeutic actions must be understood as going from rhythm to rhythm rather than from thing to thing. Which vibrations do we normally need? This question is vital, in the true sense of this word. Which vibrations die away or are aroused? Which are those to be revived or moderated? These are the therapeutic questions.

135

How though will this general view help to explain homeopathy? It is because doses are ultra-diluted that medical substances can propagate rhythms. In a massive form in fact, substance would in a way absorb its own rhythms; it would start resonating with itself, without fulfilling its role as a stimulus external to itself. It would escape indispensable destruction and fail to play with nothingness. It would reappropriate itself. Indeed, radiation physics does show that substances act above all through what is on the surface and that radiations from what lies deep are absorbed by radiant matter itself. The dilution of homeopathic matter is thus a condition of its vibratory action.

In a similar way, we can understand that the more delicate and rare are the aromas and bouquets of food and wine, the more effectively they act on our digestion. Indeed, these complex, fragile substances are easily broken down or neutralised, and easily destroyed. Now, a substance that goes back to nothingness causes radiation. 'The destructive wave' will be especially penetrating and active here. The superficial epicureanism for which smells and flavours do no more than whet the appetite must therefore, in the light of the facts, be seen as very inadequate. Pleasure has an effectiveness that goes far deeper. The question can be raised as to whether an active rhythm-analytical theory of sensation might not complete the traditional theory, wholly passive and receptive as it is. Excitation would then be a resonance that would pair with specific vibrations produced by the destruction of particular substances. All digestive values would therefore have to be transmuted. For a deep Epicureanism, ambrosia and heavenly distillations are primary necessities. When they are carefully measured out, these marvellous 'tinctures' bring us the many rare and precious essences of the plant world. They are the sources of an exhilarating kind of homeopathy and they guide us towards a sense of enhanced life. This then is the principle we should make fundamental to rhythmanalytical health: small causes have great effects; small doses have great success. We might then see the beginning of the art of micro-nutrition, if we may be allowed to use such an ugly term which does however suggest a life so joyously dematerialised! Before all else, the temporal characteristics of this micronutrition must be revealed. With a micro-foodstuff, we take in duration and rhythms rather than substance. Substance is but an opportunity for becoming; pure essence is but time that truly vibrates. We take it to be a fundamental principle that it is necessary to uphold useful and normal rhythms, to help personal rhythms and those imposed by nature to

136

harmonise, and to preserve the symphony of hormones. We must never lose sight of the fact that all exchanges take place through rhythms. Biological rhythmanalysis should undertake the task of codifying all these rhythms and of giving 'symphonic' meaning to the organic and substantial totality.

137

If diluted substances have characteristic wave effects, then the direct effect of certain waves can be very easily explained. These particular radiations can be replacements for particular substances, and Pinheiro dos Santos in fact suggests a theory of the reversibility of vibrations and vitamins:

> Some scientists, of whom Professor Centani is one . . . believe that electric charges exist in vitamins; they thus consider these to be like ions and explain their action by phenomena *which would be in biology what radiations are in physics*. Rosenkeim and Webster have shown that the action of ultraviolet radiation is similar to that of vitamin D. Ultraviolet radiation produces photons with the same frequency as those vitamin D can emit and which it has itself taken in from the sun (1, section 1, p. 26).

We shall say in passing that there is consequently a rhythmanalytical explanation for the medical action of certain insolated salts. Besides this, the highly reversible character of radiation and substance can be seen. It can therefore be maintained that certain chemical substances bring the organism not a collection of specific qualities but a group of rhythms or, as Pinheiro dos Santos has put it very well, a 'body of photons'.

Moreover, there is nothing to prevent a homeopathic substance that has taken the form of pure vibration from then being reconstituted in the form of a substance. There is in fact complete reversibility from matter to waves and from waves to matter. The role of micro-substance is perhaps very simply to initiate natural biological vibrations. It can also be explained that ultra-diluted doses are more completely preserved than massive ones because they can be restored. We arrive at the paradox that the infinitely small that is well structured and has clear rhythms is less easily lost than crude, inert matter.

138

Indeed, Pinheiro dos Santos adds to this rhythmic theory of the activities of substance an inverse hypothesis regarding the *concretion* of certain rhythms. Such for example is the curious hypothesis of the wave formation of toxins: do some cells come to receive rhythms with dangerous frequencies? There is then what he calls 'toxinic retention' (1, section 1, p. 1). Were toxins not formed that concrefy[3] and absorb harmful radiant energy, then a very minor occurrence of illness would lead to death. There follows a whole hypothesis regarding microbial relations which could form the basis of a wave bacteriology and clarify many problems. However, although Pinheiro dos Santos's explanation is coherent and rich, it cannot be seen to suggest specific experiments which would allow us to decide between the substantialist and the wave interpretation. Even so, it is in fact very important that a wave translation of classical bacteriology is possible.

Moreover, whatever laboratories decide, it will still be to Pinheiro dos Santos's credit that his thinking has shown the truly primordial character of the vibration that is fundamental to life. If inert matter has already come to terms with rhythms, then it is very certain that through its material basis, life must have profoundly rhythmic properties. It is however especially through emergence that the rhythmanalytical needs of life's process are introduced. Life is strictly contemporaneous with material transformations and impossible without their unceasing help, without the interplay of assimilation and disassimilation, and life must consequently pass through the medium of wave energy. It is only when it is dealt with statistically and globally that life seems to have temporal continuity and uniformity. From the standpoint of the elementary transformations that give rise to it, life is waves. In this respect, life is therefore directly dependent on rhythmanalysis.

Besides this, if we remember that the different kinds of matter formed by organic activity are especially complex and fragile, we shall come to consider living matter as richer in timbres, more sensitive to echoes, and more extravagant with resonance than inert matter is. Every threat of destruction, every partial death that wrecks it, the whole area of active nothingness tempting its being with a thousand intoxicating prospects are all of them opportunities for oscillations. The same is true where assimilation is concerned: every conquest of structure is accompanied by the harmonisation of many rhythms. When life is successful, it is made of well-ordered times; vertically, it is made of superimposed and richly orchestrated instants; horizon tally, it is linked to itself by the perfect cadence of successive instants that are unified in their role. We shall moreover have a better sense of the rhythmic movement of life if

139

we take it at its summits and if we study, as we shall now do, the rhythm analytical activity of the mind, that master of *arpeggio*!

III

All we have said about the wave form that the emergence of life necessarily takes could be repeated here, term for term. Indeed, conscious life is a new emergence taking place in the conditions of rarity, isolation, and unbinding that favour wave forms. In any process, the less large 140 the amount of energy involved, the clearer the wave form of exchanges of energy. Of all the different energies of life, that of the mind[4] must therefore be the closest to quantum and wave energy. It is a kind of energy in which continuity and uniformity are highly exceptional and artificial, the products of much work. The higher the psyche rises, the greater its wave movement. When we pass from the material to the mental, between matter and memory, it would be possible to set up a whole research programme allowing us to take note of the importance of the factor of repetition. Just as a heliotherapy that follows the principles of rhythmanalysis will suggest alternating periods of pigmentation and depigmentation, so a rhythmanalytical approach to teaching will establish the systematic dialectic of remembering and forgetting. We only know what we have forgotten and relearned seven times, according to indulgent teachers, the good ones in fact. However, despite these teachers' confidence in the natural reaction that prevents the mind from becoming overloaded with knowledge it cannot assimilate, they have not yet set about helping nature in this respect by contributing methods of forgetting, of 'depigmentation'. Holidays are not enough. They are too distant. They are not an integral part of the culture, of the school's temporal texture. The rhythm of school life is thus completely unbalanced; it runs counter to the elementary principles of a philosophy of repose. We must introduce oscillation into the time of work itself. Mathematics can be done to a metronome. This is a way of benefiting from the oscillations of the mind's emergence.

We shall not however prolong this discussion of the increasingly obvious wave character of the different kinds of emergence, but shall now pose a particular problem that will show the full importance of rhythmanalysis. This is the problem of the relations between psycho- 141 analysis and rhythmanalysis. More systematically than psychoanalysis does, rhythmanalysis seeks motives of duality for mental activity. It makes the same distinction between unconscious tendencies and strivings for consciousness, but it achieves a better balance than psycho analysis does between these tendencies towards opposite poles, and of the psyche's two-directional movement.

Indeed, for Pinheiro dos Santos, people can suffer from being enslaved to unconscious, confused rhythms which show a real lack of vibratory structure. Above all though, they can suffer from consciousness of their infidelity to the higher rhythms of the mind: as he says, 'human beings know they can go beyond themselves' (2, section 1, p. 5), and they both need to go beyond themselves and have a taste for doing so. Sublimation is not some deep drive; it is a call. Art is not a poor substitute for sexuality. On the contrary, sexuality is already an aesthetic tendency; it is profoundly implicated in a set of aesthetic tendencies. Pinheiro dos Santos bases his rhythmanalysis on creationist philosophy, on an active sublimation of every tendency. It is the lack of active sublimation, of sublimation with the power to draw us on, emergent and positively creationist, that unbalances psychoanalytical ambivalence and disturbs the interplay of psychic values. Not being able to *realise* an ideal love is certainly a cause of pain. Not being able to *idealise* a love we have realised is another.

Here we reach the most difficult part of Pinheiro dos Santos's theory. Let us therefore try to see how exactly creationism imposes waves of affectivity on the psyche. Should living beings seek to leave their present state and follow their own impetus, placing part of their power and energy at risk, they will at once feel the need to turn back to their acquired knowledge, back to a *support* that will ensure their impetus, as has been well shown by Jean Nogué. If on the contrary they remain at the level of what they have already acquired, then straightaway the monotonous rhythms characteristic of this state, which is closer to matter, tend increasingly to die away and the creationist reaction appears as more necessary and at the same time easier. Without this reaction, the becoming of living beings would decline into torpor. All creative evolution which is understood not in the statistical summary that is the evolution of species but in individuals, in young individuals especially, must necessarily be in wave form. Evolution in individuals is a tissue of successes and errors. The evolution of the species gives us only the sum of successes which are more or less great, more or less special, in which error is recorded in only its teratological aspects. The function of

142

the individual is on the contrary to make mistakes. If we all try out on ourselves the psychology of an attempt to create, an attempt to innovate, then however modest this attempt, or especially in fact if this attempt to create is modest, the accuracy of creationist wave psychology will be apparent. Error cannot be continued without causing harm. Success cannot be continued without risk and fragility. In its detail, the evolution of individuals takes a wave form.

From a more specifically moral standpoint, Pinheiro dos Santos is aware that repression is either liberated or corrected by cathartic method, as Freud has shown. Freud's method, though, does not go far enough: it forgets characteristics that rhythmanalysis will take great care to introduce into cathartic method. Indeed, once the repressed event has been brought to clear consciousness, it seems that for psychoanalytical theory the patient will automatically get better, that this enlightened consciousness will forgive the fault that long lay hidden, and that unconscious 'remorse' will be stilled by conscious avowal. Yet is it not to be feared that the painful process might start up again in the unconscious? Is not this painful process, as Freud himself admits, a dynamic disturbance, a disturbance of becoming rather than of a state? In order to be safe from a repetition of neurosis, for which interpretations are never in short supply, a clear system of inner forgiveness must be prepared in consciousness. We can then hope that the 'scruple' will no longer be renewed. This systematic, conscious forgiveness, set up in face of the automatic reflexes of a guilty conscience and in opposition to the dangerous slope of harmful becoming, must form the clear pole of the moral dialectic. As has often been said, psychoanalysis has underestimated the conscious, rational life of the mind. It has not seen the constant action of the mind that somehow or other always gives form to the formless and can always interpret obscure desires and instincts. Cathartic method will remain then a medical act, carried out by a skilful and knowledgeable practitioner. It is an 'operation' that may be necessary in the case of neuroses and also in the great misfortunes of criminal life. The morality that deals with detail needs a more frequent and more flexible cathartic method. It is dependent on a rhythmanalysis better suited than is psychoanalysis to following the temptations that come like waves. Moreover, when we have to attain a positive life, *inventing* good and not just doing it, only rhythmanalysis can guide us. Rhythmanalysis alone takes account of moral dualism, and Pinheiro dos Santos writes that 'the rhythmic balance of moral inflexibility and

143

kind-heartedness is the law of love and its very expression' (2, section 2, p. 12). More precisely, rhythmanalysis under the name of the
solidarity of couples has brought to light the fundamental motive of
moral duality. Since human egoism always comes back in the end to
the desire to appropriate *social* values, the seduction and conquest of
others remains the egoist's aim. The personality thus lives according to
the rhythm of conciliation and aggression 'that goes from one pole to
the other in the two contrary attitudes of the rhythm *self-love*—love of   144
others' (2, section 2, p. 6). The ambiguity of interpretations is perhaps
nowhere more visible in the closeness of its terms than in morality: all
our moral acts have a dual aim. Morality reacts on being. I respect in
order to be respected. I love in order to be loved. I do good in order to be
happy. Comparison of the self and others is the fundamental principle
of all moral proof. Of all the emotions, the moral emotion is most wave-
like. Rhythmanalytical morality sets out to regulate this wave-motion.

IV

Pinheiro dos Santos's lengthy discussions in his work have thus fur-
nished us with a number of examples of the essential polarity of mental
life that is fundamental to rhythmanalysis. In limiting ourselves in
this way, it is not possible to convey all the richness of his work. The
important point though is that we should give readers the feeling that
all life's endeavours are dialectised, that all mental activity is a passage
from one level to a higher one, and that every emergence has to have
something supporting it. Readers may find it quite easy to accept all
these polarities which are not new in philosophy, but they will no doubt
raise the following objection: how can a philosophy of time result from
these psychological and moral oppositions? Does it not appear that
duration has nothing at all to do with these problems and that all these
oppositions can be summed up by the old saying that opposites attract?

  We can answer these objections by referring to two kinds of cases
in which either the opposites confront one another in incontrovertible
hostility or else there is minimal disagreement between them. In the
first case, the duration of a state will in fact determine the intensity   145
of the opposite reaction. Politicians and teachers have often made this
observation, but it would benefit from being extended to all aspects of
life. It would then be recognised that any severe inhibition determines

accumulations of energy which must sooner or later react. The duration of a reaction that follows a long-lasting coercion is itself lengthened, hence the setting up of a rhythm that is at one and the same time both powerful and slow.

Although it would be easy to develop this point, we shall not dwell on it but shall simply ask those who criticise us to consider examples in which the opposites are less distant from each other and less hostile than those Pinheiro dos Santos examines. It will then appear that between these two fairly close poles, hesitation—that indispensable form of progress—behaves like an increasingly regular oscillation, which synchronises more and more with precise temporal rhythms. Thus, is it a matter of emotional ambivalence? Let us no longer look at irrefutably passionate and dramatic emotions. Let us consider moods that are just a little melancholy and full of fickle desires; let us consider, so to speak, temptations that do not tempt, scorn that is indulgent, kind refusal, verbal joy . . . and we shall see time begin to oscillate, all its seconds slightly contradicting and colouring each other, either dull or brilliant. Opposites unite together and then part, only to unite again, as we see in Verlaine's famous line: 'Melancholy waltz and languorous vertigo'.

Such is the minor ambivalence in which we shall see rhythmanalysis come into being. In these superficially unstable states, it is really time that offers the appropriate analytical schema; when the dialectic of consciousness and will is completely separated from interests and usefulness, it tends to become temporal. There is so little reason for *continuing* a state that the taste for breaking things off makes itself clear. Time alone is in command in this sweet, free life: everything twinkles and sparks.

If physical pain is sufficiently slight, it too comes within the competence of rhythmanalysis. With a little practice, we can for example make toothache vibrate. All we need do is calmly and attentively put it into its proper perspective and avoid the general annoyance and agitation that would *fill up* the intervals of the particular pain. The throbbing of this local pain then acquires its regular rhythm. Once this regularity has been accepted, it comes as a relief. Pain is truly restored to its local aspect because its correct temporal aspect has been fully determined.

Although we ourselves have found these detailed applications to be effective, they do require quite considerable practice. They are really only possible if the great natural rhythms upholding life have been previously restored to their importance and regularised. Here, respiration takes precedence with its slow, regular cadence which, once completely freed of all organic worries, deeply marks our temporal confidence, the confidence we have in the near future, our harmony with rhythmical time.[5]

It is regularity of breathing that a philosophy of repose must endeavour to achieve before all else. Indeed, rhythmanalysis concurs with the teachings of Indian philosophy. Romain Rolland describes to us Vivekananda's first lesson as follows:

> Learn to breathe rhythmically, in a regular way, through each nostril alternately, concentrating the mind on the nerve flow, on the centre. Add a few words to respiratory rhythm in order that it be more clearly emphasised, marked out, and directed. Let the whole body become rhythmic! Thus are learned true mastery and true repose, and calmness of both face and voice. Through rhythmic respiration everything in the organism is gradually coordinated. All the body's molecules take the same direction.[6]

147

In other words, because of their resonance regular rhythms reinforce structural symmetries. We must also stress the advice that respiratory rhythm be preserved by linking it to a slower vocal cadence. The very great effectiveness of less frequent rhythms of this kind is from our point of view essential. It shows that a low-pitched rhythm, with slow beats, can uphold and determine one that is high-pitched and with much greater frequency. Should one of life's quick rhythms be disturbed, it can be set to rights by being placed within the framework of a slower rhythm that is easier both to monitor and impose. This explains why marching along to a very discontinuous tune, with a rallying beat every two or three steps, is so beneficial for restoring calm and regularity to respiration. Too rapidly realist a conclusion would instead see the very opposite as effective, imagining that it is a rhythm with many frequencies that carries the events of a slow rhythm, as though these were additional incidents. But experience is conclusive: the mind imposes its rule on life through very few but well-chosen actions, and this is why an art of repose can be established when there is certainty of well-distributed reference points.

We shall moreover find very many confirmations of this when we study, from the standpoint of rhythmanalysis, the great sweeping rhythms that mark human life. We scarcely need to remind readers of the importance for a good and thoughtful life of living in accordance with the day's pattern, with the regular passing of the hours. Nor do we need to describe the very rhythmic duration of those who work in the fields, living in harmony with the seasons and shaping their land according to the rhythms of their labours. The importance for us, from a physical point of view, of adapting very strictly to plant rhythms has become increasingly evident since the specificity of vitamins was discovered: strawberries, peaches, and grapes have all of them their hour and are the occasion of physical renewal, in harmony with spring or autumn. The calendar of fruit follows the calendar of rhythmanalysis.

Rhythmanalysis will look anywhere and everywhere in order to discover new opportunities for creating rhythms. It firmly believes that there is a definite correspondence between natural rhythms, or alternatively that they can easily be superimposed, one rhythm imparting momentum to another. Rhythmanalysis forewarns us then of the dangers of living at odds with such rhythms, and of failing to understand our fundamental need for temporal dialectics.   148

V

We believe however that if human life is indeed placed in the framework of these natural rhythms, what we are determining is happiness, not thought. The mind needs a much closer pattern of reference points. If, as we would argue, intellectual life is to become the dominant form of life, physically speaking, with thought time prevailing over lived time, then we must devote all our efforts to the quest for an active repose that finds no satisfaction in what is freely bestowed by the hour and the season. It would seem that for Pinheiro dos Santos this active, vibrating repose corresponds to the lyric state. The Brazilian philosopher has close knowledge of modern French literature, in particular of Valéry and Claudel, whom he greatly admires. He submits to each in turn, to the power and the rhetoric of Claudel's writing, and then to the subtle ambiguities of Paul Valéry's thought. In Valéry, he appreciates most of all the supreme art of the poet as, skilfully, he disturbs our calm and calms our disturbance and moves from our heart to our mind, only to return at once from mind to heart.

Yet Pinheiro dos Santos does not rest content with this rather coldly intellectual interpretation of the lyric life. He prefers that lyricism should continue to be regarded as a purely physical charm, a myth that lulls us to sleep, a complex binding us to our past, to our youth and its impetuosity. Indeed, he suggests a lyric myth for rhythmanalysis which could well be called the Orpheus complex. This complex would correspond to our first and fundamental need to give pleasure and to offer solace; it would be revealed in the caresses of tender sympathy, and characterised by the attitude in which our being gains pleasure through the giving of pleasure, by the attitude of making some kind of offering. The Orpheus complex would thus be the antithesis of the Oedipus complex. Poetic interpretations of this Orpheus complex may be seen in Rilke's orphic lyricism, as Félix Bertaux has called it,[7] a lyricism which egotistically lives out an indeterminate love of others. How very sweet it is to love anyone or anything, indiscriminately! How delightful ever to live at the moment of falling in love, ever amidst love's first rapturous declarations! This then is the basis of a theory of formal pleasure which is the very opposite of the theory of that immediately objective material pleasure, which in the Oedipus complex binds the unfortunate child to the face that is first seen above the cradle. Rhythmanalysis is the complete antithesis of psychoanalysis in that it is a theory of childhood rediscovered, of childhood which remains a possibility for us always, always opening a limitless future to our dreams. It is interesting to note here that Pinheiro dos Santos has, in an essay in which he takes issue with Freud's work on Leonardo da Vinci, set out to explain this artist's creative genius in terms of an eternal childhood. Creationism is in fact nothing other than the process of growing perpetually younger, a method of systematic wonderment that rediscovers a pair of wondering eyes with which to look upon familiar sights. Every lyric state must originate in this truly enthusiastic knowledge. The child is our master, as Pope once said.[8] Childhood is the source of all our rhythms and it is in childhood that these rhythms are creative and formative. Adults must be rhythmanalysed in order that they may be restored to the discipline of that rhythmic activity to which they owe their own youth and its development.

VI

We ourselves would prefer that the lyric state be subjected to some kind of mental elaboration, thus setting ourselves at some distance from the unconscious powers that imprison us in the Orpheus complex. We have,

for this reason, turned our attention to the uppermost regions of superimposed time, to thought time, in our search for the most clear-cut and
therefore most invigorating of dialectics. We have, for example, sought   150
to experience all Valéry's poetry in our own way by applying to these
poems the structures implicit in the dialectic of time. This may well be
too abstract and too personal an approach, suggested all too readily by
habits bred of dry and dusty philosophy. Yet we have discovered that
as a result of using this method, which in effect impoverishes, we can
hear rare and precious echoes; we have experienced in particular how
the temporal structure found in ambiguity can help us to intellectualise
rhythms produced by sound, and so enable us to *think* that poetry which
will not reveal all its charms when we confine ourselves to speaking or
feeling it. We have come to realise that it is the idea that sings its song,
that the complex interplay of ideas has its own tonality, a tonality that
can call forth deep within us all a faint, soft murmuring. If we speak
soundlessly and allow image to follow image in quick succession, so
that we are living at the meeting point, the point of superimposition,
of all the different interpretations, we understand the nature of a truly
mental, truly intellectual, lyric state. Reality is enfolded and adorned
by the rich garment of conditionals. In place of the association of ideas
there comes the ever possible dissociation of interpretations. The mind
takes pleasure in its refusal of all that it once found unfailingly attractive: it discovers all the delights of poetry in its destruction of poetry,
as it contradicts the sweet spring and resists all charming things. This, it
must be said, is a highly epicurean asceticism, since in this conditional
form pleasure seems more vibrant. Poetry is thus freed from the rules of
habit, to become once again the model of rhythmic life and thought that
it used to be, and so it offers us the best possible way of rhythmanalys-
ing our mental life, in order that the mind may regain its mastery of all
the dialectics of duration.

NOTES

1. Bachelard's footnote: L. Pinheiro dos Santos teaches philosophy at the
University of Porto in Brazil; the work referred to is *La Rythmanalyse*, Société
de psychologie et de philosophie, Rio de Janeiro (1931). This work was not
published commercially and has not been traceable, despite the efforts of many
Bachelardian scholars over the years: it appears to have been sent privately to

Index

About the Author

Gaston Bachelard (1884–1962) was Professor of Philosophy at the University of Dijon, and later held the Chair of History of Philosophy of Science at La Sorbonne. His influence is attested by thinkers as diverse as Derrida, Foucault and Barthes.